WORKERS' INFERNO

RAMSINA LEE

PRAISE FOR *WORKERS' INFERNO*

Ramsina Lee, in this book has possibly for the first time not only described the events of the day but also meticulously documented in narrative form, the impact on the lives of some of those that were involved in one of Australia's most prominent industrial disasters. The loss and suffering of the Longford workers and their families was profound. Only now, twenty years after the terror and trauma of that day has the subsequent loss and suffering been accurately portrayed.

Jim Ward
Longford Survivor

Ramsina Lee narrates the story of the workers who were on the shift during the tragic explosion at Esso Longford on 25 September 1998 including my late father John Lowery who was killed during the explosion. Her book describes the lives of regular loving families whose world was shattered because of the explosion and the subsequent further trauma of the harrowing experiences of the events that followed. It is a book that I will treasure and pass on to my children in remembrance of my father who died unnecessarily and could now have been enjoying retirement and my family. I hope this book is an impetus for major positive change in the area of work health & safety throughout all workplaces.

Shannon Lyons
Daughter of the late John Lowery

Ramsina Lee has very thoughtfully allowed the families of those affected by the Longford gas plant explosion to have a say. She tastefully weaves created scenes and settings with interviews of those directly involved, and their family members. The catastrophic personal events and tragedies of 25 September 1998, the immediate aftermath and the long lasting effects are felt in the heartfelt interviews. The intense love for lives lost and lives forever changed. The callousness of the big corporation and the blatant disregard for the 'scapegoat' and their family. A champion for workplace safety and the preciousness of life and love.

Elizabeth Coleman Gray

In *Workers' Inferno* Ramsina Lee has narrated the story of the events preceding, during and after the explosion at the Esso gas plant on 25 September 1998. My late father Peter Wilson was one of the two workers who were killed in the catastrophe. Whilst the surviving workers, their families and the families of the deceased workers experienced the poignant reality of the deaths and injuries that resulted because of the tragedy, there has not been an understanding or visibility by the community at large of the events of the day or the impact it had on the families. Ramsina Lee provides that missing piece. This book will form a part of the Wilson family history for my children and the children of my brothers Brett & Luc; lest our father's story be forgotten.

James Wilson
Son of the late Peter Wilson

This book is a gripping and emotion-laden account of the tragedy that occurred at the Esso gas plant at Longford, Victoria, in September 1998. Ramsina Lee begins by delving into the happy but sometimes complicated family lives of many of the workers involved in the explosion. Then she describes the confusion, bewilderment and heroism of these workers as the breakdown of the gas plant causes the explosions and fires that rip through the workplace and change their lives forever. Finally, the emotional aftermath of the tragedy tears families apart, leaving a permanent human cost to be borne. I read *Workers' Inferno* in one sitting, and felt the fears, emotions and bitterness of the workers and their families as an avoidable workplace tragedy that changed the course of human lives.

The message of the book is clear — take workplace health, safety and training seriously so that workers can return home safely to their loved ones at the end of each working day.

Joe de Bruyn
National President,
Shop, Distributive & Allied Employees' Association

Among other things this is the story of Jim Ward, the operator who Esso blamed for the disaster at the Longford Royal Commission. Esso's tactic was patently absurd and the public reaction was so intense, that Esso dared not blame Ward at the subsequent jury trial. The company's explanation was a classic case of blaming the worker for what the jury found was actually a case of corporate crime. *Workers' Inferno* is a beautifully written, almost poetic account - passionate, compassionate engrossing and unforgettable.

Andrew Hopkins
Emeritus Professor of Sociology
Australian National University
author of Lessons from Longford: The Esso Gas Plant Explosion

What can I say? Ramsina Lee has provided an account of an event about which I knew so little. Not just describing the horror of the events of 25 September 1998 at Longford, but more importantly bringing life, character and spirit to the men and families whose lives were irrevocably changed on that day. This is not an easy story to sit with, but it is compelling.

Jon Graham
Clinical Director,
Institute of Specialist Dispute Resolution (ISDR)
Investigator of Workplace Psychological Violence

Once in a while, a book comes along that is hard to put down. *Workers' Inferno* is such a book. Ramsina Lee successfully tells the story of everyday workers in Australia, behaving responsibly in their everyday work, then suddenly finding themselves in an unprecedented situation, far beyond what they have been trained for, desperately trying to shut down a gas processing plant out of control. All the while feeling as if they might not survive. We meet them and their families as they cope with this disaster and its aftermath including the death of two workmates. It is sobering. This is a book for many — blue collar workers, management and unions; most of all it is a reminder of the necessity to respect human life, in all aspects of life.

I unconditionally recommend this book to read as a milestone in the understanding of these areas.

Kathleen Prendergast

Published in Australia by Significance Publishing
Postal: PO Box 1799, Neutral Bay, NSW 2089
Tel: +61 411 282 350
Email: info@ramsinalee.com
Website: www.ramsinalee.com

First published in Australia 2018
Copyright © Ramsina Lee 2018

All rights reserved. No part of this publication may be reproduced, stored in a retrieval system, or transmitted, in any form or by any means without the prior written permission of the publisher, nor be otherwise circulated in any form of binding or cover other than that in which it is published and without a similar condition being imposed on the subsequent purchaser.

National Library of Australia Cataloguing in Publication entry

A catalogue record for this book is available from the National Library of Australia

ISBN: 978-0-6483100-0-6 (paperback)
ISBN: 978-0-6483100-2-0 (hardback)
ISBN: 978-0-6483100-3-7 (epub)

Cover photography by pixabay.com and shutterstock.com
Cover layout and design by @difrats
Book typesetting by Sophie White

Printed by Ingram Spark

Disclaimer
All care has been taken in the preparation of the information herein, but no responsibility can be accepted by the publisher or author for any damages resulting from the misinterpretation of this work. All contact details given in this book were current at the time of publication, but are subject to change.

The advice given in this book is based on the experience of the individuals. Professionals should be consulted for individual problems. The author and publisher shall not be responsible for any person with regard to any loss or damage caused directly or indirectly by the information in this book.

The author has attempted to recreate events, locales and conversations from interviews conducted in 1999 and in 2018. In order to maintain anonymity, in some instances the names of individuals and places, identifying characteristics and details including physical properties, occupations and places of residence have been changed.

The author acknowledges the traditional custodians of the land this book has been prepared on and will travel to across Australia. The author pays respect to Elders past and present, and acknowledges all Aboriginal and Torres Strait Islander workers.

DEDICATION

In memory of Peter Wilson and John Lowery who died during the explosion on 25 September 1998. And in memory of the late George Parker, the union delegate who emerged as the rock and the binding force during the events after the explosion.

To all the medical and emergency personnel and the community at large that responded to the catastrophe. You are the unsung heroes.

In memory of my late father Babajan Danyel — the worker who toiled without a union. My hero. Your commitment to your family and marriage compelled you to risk your life during the terrorising, harsh and dehumanising conditions of being illegally shipped to foreign lands to find work. You risked your life for us. You taught us that work, human dignity and self-respect are intertwined with the human spirit. So, it is a human right to have dignified work and to be paid for that work. One day, I will tell the world your story. I miss you immensely. But I know you're hard at work in heaven as you were here on Earth, making sure we're looked after. Thank you!

CONTENTS

Author's Note		11
Foreword		15
Prologue		19
Longford Lament		23
Introduction		25
Chapter One	The Football Match and the Ward Family	27
Chapter Two	The Union Meeting	33
Chapter Three	Bill Shorten is Elected Union Secretary	37
Chapter Four	Ronnie Rawson	41
Chapter Five	Robert Miller	45
Chapter Six	Peter Wilson	47
Chapter Seven	John Lowery the Brother	51
Chapter Eight	John Lowery the In-Law	57
Chapter Nine	John Lowery, Shannon and Ashley	63
Chapter Ten	Marty and Sue-Ellen Jackson	69
Chapter Eleven	25 September 1998 Trouble at the Plant	73
Chapter Twelve	Bill Shorten Hears the News	105
Chapter Thirteen	George Parker and the Holiday	107
Chapter Fourteen	Luc Wilson	111
Chapter Fifteen	Locky Wilson	115
Chapter Sixteen	Elizabeth, Sue-Ellen and Rhonda Hear the News	121
Chapter Seventeen	More Explosions at the Plant	125
Chapter Eighteen	News about John and Peter	139
Chapter Nineteen	Two Funerals	147
Chapter Twenty	The Longford Royal Commission	153
Chapter Twenty-One	The Ward Story Continues	161
Epilogue		173
How It All Ends		181
Appendices		183
Footnotes		193
About the Author		195
Acknowledgements		197

AUTHOR'S NOTE

Let me say at the start, this book is not about Esso. It is the story of the people whose lives were changed as a result of a tragic event that took place at the Esso gas plant in Longford, Victoria. The purpose of telling the story is twofold. First and foremost, it is to give voice to the voiceless entangled in a system that is focussed predominantly on the legal outcomes of such events. Secondly, it is to inspire collegial and collaborative relations in workplaces by bringing the human element of such circumstances into the focus of decision makers and influencers in organisations.

While industry and labour are disconnected and work against each other, the story of the Esso Longford disaster will repeat itself in other places, through other people and in different forms.

However, when industry and labour work together and in synergy, the success will be as the murmuration of a flock of birds, bringing order in a multi-agent social system. Let's go for murmuration!

I have one request of you. Please read my book.

It might be obvious that as an author who has researched, transcribed, narrated and nurtured my book for years, a book I am very passionate about, I would ask people to read it. But I'm not asking you to read through it as you would a novel or a story of interest. I'm not asking you to skim through it. I'm asking you to read my book with all your senses engaged and I want to transport you, through your imagination, to the scenes and events I have described. I want you to feel the sense of belonging and mateship on the football field in Maffra. I want you to join in the family connections and love within the families described throughout the chapters. Then, I want you to be transported to the gas plant where things were different, very different to every other day. I want you to feel the confusion and uncertainty of what was happening in the plant on the morning of the 25 September 1998. Then I want you to feel the trembling caused by successive explosions. Hear the blaring sirens and alarms, see the monstrous black smoke clouds and smell the vapours that assaulted the lungs of the workers. See your colleagues blackened by the effects of the blasts around you. Feel their shock as they tried to figure out what was happening. Feel the grief of the workers and their families. The grief for the loss of lives of family and friends and grief for shattered dreams. I want you to hear your name mentioned during the Longford Royal Commission repeatedly by your employer in a campaign to blame you for the explosion.

I want you to read my book in this way because when you do, you will come to two conclusions.

The first conclusion is that the life, safety and welfare of workers are not collateral. They do not belong on the negotiation tables of enterprise bargaining or any other negotiations; they are not bargaining chips. The life, safety and welfare of workers are sacred. And they are to be revered as sacred. For, when they are revered as sacred they will receive the respect and priority they deserve. The life, safety and welfare of workers have to be at the centre of every decision, every change, and workers need to be consulted in situations and decisions where their life, safety and welfare may be affected.

The second conclusion you will come to is the importance of the freedom of association and protection of the right to organise. What does that mean? It means that every person has the right to come together with other people to collectively express, promote, pursue and defend their common interests. In this case, it means the right for workers to come together with other workers to collectively express, promote, pursue and defend their common interests at work, and those interests affecting their work.

This right is so important that the United Nations declared it a fundamental human right in 1948. In Australia, we have taken this human right for granted for too long. And in recent decades, this right has been under attack. This has resulted in the denigration of worker unity and therefore, the successful progressive and persistent erosion of the working conditions of workers. Especially their safety. The time has come for society to recognise the important role unions have in workplaces and allow their rightful place and voice in representing their members.

Conversely, the time has come for unions to make their politics second order and lead with the interests of their members as their foremost goal.

FOREWORD

Workers' Inferno is published on the occasion of the 20th anniversary of the gas plant explosion at Longford near Melbourne, an explosion that killed two workers and injured several others. Longford is one of the iconic accidents in the oil and gas industry. The lessons from Longford were widely discussed in industry circles around the world. That is not to say that the industry learnt from them, because such accidents continue to happen with alarming regularity. My own book focused on these lessons. The human tragedy formed the taken-for-granted backdrop. Ramsina Lee's timely book redresses that balance. It tells the story of the men and women whose lives were destroyed that day, either literally or figuratively. It tells the story in their own words. The result is a stunning piece of oral labour history.

Among other things, this is the story of Jim Ward, the operator who Esso blamed for the disaster at the Royal Commission. Esso's tactic was patently absurd and the public reaction was so intense that Esso dared not blame Ward at the subsequent jury trial. The company's explanation was a classic case of blaming the worker for what the jury found was actually a case of corporate crime.

Workers' Inferno is a beautifully written, almost poetic account — passionate, compassionate engrossing and unforgettable.

Andrew Hopkins
Emeritus Professor of Sociology
Australian National University
author of Lessons from Longford: The Esso Gas Plant Explosion

------

On 25 September 1998, two good men, Peter Wilson and John Lowery were killed in the Longford gas plant explosion.

Eight other workers were injured. Some permanently. Many others suffered psychological trauma that is still with them to this day. As a result of the explosion, the state of Victoria in southern Australia lost its supply of gas for two weeks. The economic consequences were colossal.

The operator of the gas plant, Esso Australia, a subsidiary of Exxon Mobil subsequently blamed its workers for the explosion in a Royal Commission inquiry.

The findings of that inquiry exonerated the workers and laid blame with Esso.

Esso was subsequently prosecuted in the Supreme Court of Victoria and was found guilty of eleven breaches of the state occupational health and safety legislation. It was fined $AUD2.75 million. Twenty years later this fine is still the largest fine for a breach of an occupational health and safety legislation in Australia.

Approximately one year after the explosion, in October 1999, Ramsina Lee, a woman with a mind imbued with curiosity and a heart filled with empathy travelled with her husband Michael, some two-and-a-half hours to a little town in Gippsland to meet with workers who were at Longford on that fateful day.

Ramsina sat with the workers, asked questions and listened. She decided that she wanted to tell their story.

Life often gets in the way of plans. Ramsina's desire to tell the Longford workers' stories became waylaid by raising a family and pursuing a career...however she never forgot the voices she heard on that day she travelled to Gippsland.

Almost twenty years have passed since Ramsina recorded those stories. The confluence of a significant milestone anniversary and an abiding desire to document the tragedy of Longford had never completely subsided. This book is a testament to Ramsina's commitment and determination to honour her promise to herself... to tell the Longford workers' stories.

The history of industrial disasters is well documented. Numerous inquiries have led to careful analysis, cause and effect, findings and recommendations and ultimately...blame. Yet history continues to repeat.

So often the human story associated with such events is suffocated by technical facts, argument and the focus on liability.

Ramsina Lee, in this book has possibly for the first time not only described the events of the day but also meticulously documented in narrative form, the impact on the lives of some of those that were involved in one of Australia's most prominent industrial disasters.

The loss and suffering of the Longford workers and their families was profound.

Only now, twenty years after the terror and trauma of that day has the subsequent loss and suffering been accurately portrayed.

Lives were changed forever on that day through no fault of their own. It is so often said by victims that all they seek is 'closure'. The truth is though, that closure is never attained. There is no cauterisation of the wound that is trauma.

You just live with it as best you can.

There can be no doubt that Esso Australia is a rogue employer. There is too much evidence to suggest otherwise.

Workers' Inferno is a valuable record of the impact of a rogue employer on ordinary peoples' lives.

Let this be a history lesson.

Jim Ward
Longford Survivor

PROLOGUE

Three weeks before our wedding, on a cold spring day in September of 1998, I was in Melbourne at my fiancé's home in the suburb of Strathmore. For me, getting married was an exhilarating time after such a long time of independence as an unmarried and unattached career woman. One side of me wanted so desperately to marry Michael and finally settle down and start a family. The other side of me was anxious, wondering what I would have to give up. Above all, I hoped our marriage would last.

As I sat in the small, sunny kitchen of Michael's Victorian townhouse pondering these issues, my attention was drawn to a television news broadcast. I wasn't sure if I had heard correctly. Something about an explosion at a gas plant in Victoria. All gas supplies in Victoria were affected. The information in the news flash was so sketchy. It ended as abruptly as it had started.

The strangest thing happened to me: I sensed myself functioning in my heart and in my head simultaneously. My head was thinking about the immediate practical consequences, like how we would shower or cook without gas supply to our home, yet in my heart I felt compassion, even grief, for the people involved. I wondered if there were any fatalities — and how were their families? I felt frustrated due to the lack of further information.

For some reason, I was drawn into this event. I felt anger towards the system and anxiety for the employees and their families. This anger frightened me because I didn't know from where the intensity of these emotions originated. I called Michael at his work.

I asked him if he had any news on what had happened. He repeated what I already knew. Towards the end of our conversation he added, "The most recent news is that some of the workers might have been killed."

After ending the phone call, I sat on the leather sofa in the lounge room and wept. I felt deep compassion and sympathy for these workers and their families. It was as if this tragedy had somehow tapped into my past. I called Michael back. I told him that I didn't know why, but I needed to reach out to this group of workers and to hear their story.

"I can't see how that's going to happen," he said. "They'll be caught up with all sorts of legal investigations. How do we even get their contact details?"

I didn't know the answers to these rational questions. I just knew I had to meet with the workers and hear their stories — directly from them. After a long silence, Michael continued, "but if you can manage a meeting, I'll take time off work and we can drive down."

I spent the next few days listening to the news and piecing together as many of the facts as I could. I felt preoccupied by this event. Michael and I were going to be married in three weeks' time and yet my attention was more on this event than our wedding. Admittedly, everything for the wedding was planned and scheduled so there wasn't much for me to do. But it seemed extraordinary to me that I was more focused on the explosion at Esso than on my own wedding.

Twelve months after the explosion at Esso Longford, Michael and I had married, relocated to Strathmore and started a new life. With Michael's support, and a recommendation from the then-secretary of the Australian Council of Trade Unions (ACTU) Bill Kelty, I had started researching case studies for a book I was preparing to write. I had met and interviewed many extraordinary people from all around Australia during that period. However, I felt pulled to the Esso Longford explosion more than any other case. I felt engrossed with the court proceedings and the outcomes of many complex and long-standing issues that stemmed from them. I was mostly drawn to the employees who were on site on the day of the explosion. I wanted to include them in my writings.

Finally, in the second weekend of October 1999, with assistance from the local organiser of the Australian Workers Union, Terry Lee, I had meetings arranged with some of the workers who were on the shift when the explosion happened. Michael and I drove to Stratford, to the home of the union shop steward George Parker and his wife Judy, to meet and speak with the men who were on the shift that had changed their lives forever.

The people we met were extraordinary. I felt an immediate connection with their sense of community, deep and profound sadness, innocence — even naivety. Most of all, I empathised with their grief.

After a full day of interviewing and recording the workers and some of their families, Michael and I drove back to Strathmore; each of us affected in a different way by the people we had met and the horror of their story. With hearts filled with compassion and minds filled with images of a catastrophe we could not fathom, we drove home in silence. That night, I recorded the following account in my journal:

"... hearing the testimonies of the men who were on site during that fateful day was very moving. At one stage, I didn't think I could listen to the sadness and see the men cry any longer. In fact, unable to sleep, I was so stirred that my only comfort was in writing a poem I called the Longford Lament."

LONGFORD LAMENT

On a crisp spring day at Longford, the men were geared to go. Another day of working for big corporate Esso.

Proud of the company they thought was their mate, they did their job getting gas to the State.

Doing their rounds they said 'hello', shared a joke or two. They talked about sports, their kids 'n' wives, and other mates they knew.

The handover differed today, there was a glitch or two. But, of the terror to come, no one knew.

One by one, their troubles unfolded. Bit by bit they made their approach. Gas leaked, temperatures dropped, their efforts in vain, were beyond reproach.

All of a sudden, the nightmare began. Explosions! Fire! Soot and smoke! "Where's my dad?" asked Luc, son of Pete — a caring mate and a great bloke.

Concealed in the moment what lay ahead. Sadness, betrayal and the death of two. In the pain and confusion, only mateship and union would see them through.

In the quiet and calm they came in droves. Prodding, asking, and searching, for an answer ... a clue. Without care or concern for the workers, or Peter and John — the death of two.

"You fellas did a great job," the bosses said. But in the courtroom they turned and blamed them instead.

Judas reincarnated on that sad and dreary day. When all they did was a fair day's work for a fair day's pay.

Feeling betrayed, defeated, helpless and small, their spirits low and future grim. First victims, now felons the dice were cast, the blame was on Ron and Jim.

In fear and confusion they sought help for their case. They turned to the union to change their fate. It came to their rescue ... that noble mace.

Victory was theirs on judgment day. The accusers were stunned! Courage, faith in union and mates, with God saw justice done.

This journey of grief is a warning for all. Take heed workers everywhere for united we stand, divided we fall. Of split and rift always beware.

Stand tall, stand strong, stand united in all. And when trouble comes trust in God above all.

© *Ramsina Lee 1999*

Writing the poem was cathartic. I felt that I had transferred my feelings, especially of overwhelming helplessness, to the poem.

Then came the harrowing task of transcribing everything I had recorded on audio cassette tapes. Listening to the recordings was harder than hearing it for the first time. I felt greater anguish because I knew the stories that would unfold.

INTRODUCTION

This book is based on events that took place at an Esso operation (a subsidiary of Exxon Corporation) at Longford, Victoria, Australia on the morning shift of Friday 25 September 1998.

During this particular shift, numerous maintenance personnel, administration staff and fifteen shift workers were on site at the Esso gas production and processing plant when a vital piece of processing equipment exploded and caught fire. Two workers were killed and a number of workers were injured.

The initial explosion caused numerous other explosions and fires, which took two days to extinguish. Through a proclamation by the State Governor, all gas supply, except for critical areas of the State, was shut off.

Three weeks after the explosion, the Governor of the State of Victoria established the Longford Royal Commission to investigate and report on the causes of the explosion and fire which occurred on 25 September 1998, and the failure of gas supply from the Longford facilities following that explosion and fire. In addition, to investigate and report on whether issues relating to the design of the plant and numerous operating and maintenance procedures followed by Esso caused or contributed to the fire and explosion and failure of gas supply; and future steps to be taken to avoid similar occurrences from taking place.[1]

The Commission proceedings started in November and lasted four months.

Thirteen parties, which were concerned by the outcome of the Royal Commission, were allowed to appear before it and make submissions. They included entities such as the company and its associates, suppliers of chemicals, insurance groups and the trade union movement. One trade union in particular, the Australian Workers' Union (Victorian Branch), represented the interests of the plant operators who were directly connected with the explosion on the day.

CHAPTER ONE

THE FOOTBALL MATCH AND THE WARD FAMILY

"THERE IS NOTHING ON THIS EARTH MORE PRIZED THAN TRUE FRIENDSHIP."

Thomas Aquinas

It was a bright, warm sunny day in Maffra, a small attractive town lined with trees and buildings from the nineteenth century, a short distance out of Sale, eastern Victoria. A crowd had gathered on the field to watch a friendly Australian Rules Football match between two local teams. The crowd included some locals, Peter, whose son Luc was playing, and some of the workers from Esso who were there to watch Luc play. Another mate Marty, who had recently joined these men as an operator at Esso, was coaching Luc's team.

The game was played hard by both teams. Luc kicked three goals consecutively. Spectators cheered and applauded him.

One of the locals turned to Peter and laughingly asked, "How's that for a friendly game? He should be playing up in the city, mate!"

Peter nodded and said, "Yeah, but he doesn't wanna leave town. He says he doesn't want to live in the big smoke. He wants to live in Sale and enjoy the country life he's grown up in."

"I don't blame him. No one in their right mind could leave the fresh air, the community spirit and the mateship. I wouldn't leave it for the world," replied the local.

When the game ended and spectators and friends dispersed, the community spirit remained. Some of the spectators ruffled Luc's hair affectionately and congratulated him. Others walked past and waved, as if to a local hero. Peter walked up to his son and patted Luc's back and softly punched his gut. "That's my son," he teased with pride in his voice. They laughed and walked to the car.

"A HAPPY FAMILY IS BUT AN EARLIER HEAVEN"

George Bernard Shaw

Jim Ward drove through the main street of his hometown. The familiar sight of the small shopping strip, the two pubs (the top pub and the bottom pub) and the ever-busy general store seemed to never change, he thought.

This town was socially conservative, working-class, decent and, above all else, community-minded. He loved its inhabitants and particularly one local girl: the girl he married when they were twenty-three years old.

As he pulled into the driveway of his home, a modest timber cottage overlooking a lake, he felt gratitude for his life. He'd been married to Elizabeth, his soulmate, for eighteen years and they had two children that he absolutely adored. He turned off the engine and sat in the car for a while and wondered if life could get any better.

When he walked into the house, he followed his daily routine. First, he walked to the kitchen where Elizabeth was preparing dinner. He kissed her affectionately on the forehead. She kissed him back.

Elizabeth smiled and continued preparing the meal. Before he could exit the kitchen, his children, ten-year-old Haydn and eight-year-old Katlyn came sprinting down the hallway and into the kitchen, leaping onto their dad and clinging to any available limb whilst shrieking and laughing.

The usual playful frolicking ensued ... Haydn desperately trying to wrestle his dad to the floor by clinging onto his safety-boot clad feet while Katlyn, like a koala in a tree, perched on his back, wrapped her legs around his waist and arms around his neck, urging him to give her a "pony ride".

Things would inevitably calm down as soon as Elizabeth set the table and plated up the meals.

During dinner, Jim's charisma was the centre of the family banter. "So, Katie, what was your magic moment today?" Jim asked.

Katlyn, full of excitement and enthusiasm to share the highlights of her day, dived into the discussion. "I came first in Maths, Dad! And the funny thing is I spent more time practising my high jumps than studying for Maths."

While the family laughed at this response, she took a quick, shallow breath to avoid being interrupted and asked, "Dad, can we practise netball after dinner?"

Jim smiled and said, "No worries, possum."

They loved these times of father-daughter togetherness. Neither would ever say as much but the mutual affection was palpable.

Haydn and Elizabeth smiled and rolled their eyes in pleasure at seeing Jim and Katlyn's close bond dominate the family conversation. Then Jim turned to Haydn and asked, "What about you mate, did you have a magic moment today?"

Haydn thought for a short while and with a smile replied, "Yep ... me and Grandpa got the old stationary engine running after school today. Started first pull. Grandpa thought it'd never start and that it was seized but I could turn it over with one hand and it had good compression."

Elizabeth smiled proudly. The bond between her father and Haydn and their mutual love of historical motors meant so much to her. She loved the fact that her first-born had developed a passion for mechanics and machinery. She'd grown up with it and felt comfortable talking about such a male-dominated interest.

Haydn looked at his mother and asked, "What about you, Mum?"

Elizabeth wrinkled her nose and replied, "Well, as the Deputy Principal, I'm always happy when no one gets sent to my office for detention. So that was my magic moment: well-behaved students." Everyone laughed.

Elizabeth turned to Jim and asked, "What about you honey, what was your magic moment?"

Jim, studying his plate, formed a witty reply to make the family laugh, "Well, my magic moment was walking out of the gas plant today knowing that I won't be back there for six whole days ... ahhh the beauty of shiftwork." He raised his beer as if to toast everyone and they all responded with imaginary glasses.

Everyone laughed and cheered.

Elizabeth looked at her family and felt a deep sense of satisfaction and gratitude for how her life had turned out. She felt particularly peaceful about her marriage to Jim. Their marriage was rock solid. She was confident that there was nothing in the world that could ruin it.

After dinner, Jim looked at Elizabeth and said, "Liz, I love you and I love your cooking. What I love most about you is that you and Haydn will clear the dinner table while Katy and I go train the player of the match for this Saturday!" Katlyn knew this was her cue to get away before the cleaning chores started.

Jim and Katlyn sprang up off their chairs. Katlyn cheerfully yelled, "Race you, Dad!" as she ran to the backyard, laughing.

Jim followed in the same happy spirit, laughing and yelling, "Well, that's not fair, you had a head start ... and you're much younger than me!"

Haydn looked at Elizabeth who was smiling at him, conveying her contentment in life.

That evening, everyone had settled back into the house for the evening routines of homework, playtime and the evening television. A news item announced an oil rig explosion in South Africa, where seven people were killed.

Suddenly, Katlyn started sobbing. Jim leant over the arm of his lounge chair and looked at Katlyn sitting quietly sobbing on the floor. He glanced at Elizabeth quizzically before asking Katlyn "What's wrong, baby?"

Then he got up off the lounge chair, gently knelt next to Katlyn and, while holding her, he asked again, "What's the matter, possum?"

Finally, Katlyn stopped sobbing, and looking into Jim's face, she asked, "Dad, some people died in an oil rig. What if that happens to you at the gas plant?" Then, as if experiencing a premonition, she started to sob again.

Jim slowly and gently stroked Katlyn's long blonde hair, picked her up and cuddled her, "Baby, that could *never* happen to me. The gas plant is safe. Please don't be upset," he said, dabbing at the tears running down her face.

Katlyn continued to sob quietly as she hugged Jim. "I hope not, Dad. It's so scary," she said gently, whispering through her tears.

CHAPTER TWO

THE UNION MEETING

> "IT IS ESSENTIAL THAT THERE SHOULD BE ORGANIZATION OF LABOR. THIS IS AN ERA OF ORGANIZATION. CAPITAL ORGANIZES AND THEREFORE LABOR MUST ORGANIZE ."
>
> Theodore Roosevelt

The Esso operation at Longford was situated twenty kilometres from Sale, in South Gippsland, Victoria. It was about a two-and-a-half hour drive from Melbourne to Sale. The operation had been supplying most of Victoria's gas needs since 1969. It occupied 169 hectares of land, comprised three gas plants and one crude oil stabilisation plant. The aerial view of the property looked more like a gated suburb, occupying two sides of a section along Garrets Road. On one side were the parking and helipads and on the other side was the facility which was organised into sections divided by roads. Names such as Crude Centre Road, Gas Plant One Control Room Road, Fire Shed Road and Slugcatchers Road reflected some of the processes and locations of the giant machinery used for the processes.

The facility adjoined farmland but was remote from residential areas. As one of Australia's most important industrial facilities, its operations were continuous and needed to be staffed twenty-four hours, every day of the year. This had been the onshore receiving point for oil and gas output from Bass Strait for decades.

The population of Gippsland had benefited immensely from the operation of the plant in their nearby towns. Esso had provided work for generations of workers who were content and proud to be associated with this giant organisation. Before Apple and Google even existed, Esso had dominated the world corporate stage for branding and employment. This made the task of attracting deep allegiances with the local trade unions very difficult.

On a cold frosty morning, a meeting of the members of the Australian Workers' Union was underway in the canteen before the morning shift started. The workers sat comfortably on plastic café chairs that surrounded round wooden dinner tables scattered throughout the canteen. They drank their hot tea or coffee and ate steaming meat pies, cakes or homemade snacks as they talked and jibed with one another.

Terry, the union organiser, stood up in front of the lunchroom to address the friendly group. He had known these men as long as he had been an organiser with the union. He felt a sense of kinship with them after years of advocating on their behalf and negotiating their interests before Esso management.

Clearing his throat, Terry called for attention. The men responded, gradually ceasing the talk and banter. He continued, "Men, as you know, Judd was the shop steward of the union for quite some time and he'd been doing a great job. But the bosses saw that, so they promoted him."

The men nodded and smiled, while a few of them hissed jokingly and laughed. Terry joined the laughter before continuing. "So now it's time for you to elect a new shop steward." The silence that followed indicated this was not going to be a highly contested position. Terry continued as he looked around the lunchroom, "Do I have any nominations?"

Everyone was uncomfortably silent. Then Alex, one of the quieter men, spoke up. "Mate, why do we need a union? I mean, I don't have any problems with unions. I know working for this company is hard work but it's also a privilege for most of us. They pay well, we get good rosters, the shifts suit us fine and we work with the best men in Gippsland."

One of the other men objected and asked, "Yeah, mate, but who got us those conditions to start with?"

The rest of the men started to talk at the same time. "Yep, that's right! Remember when we used to get our petrol allowance and safety awards? What happened to them? You can't trust the bastards!" referring to management. "Yeah, they're bad today and worse tomorrow!" The commotion continued for a short while, followed by silence.

Barry, who was the joker in the group, stood up and flippantly addressed the crowd, imitating an English nobleman, "Silence ... silence ... gentlemen. It is my pleasure to nominate my esteemed colleague Charlie for the position of shop steward. After all, he doesn't seem to do much around here."

The lunchroom immediately filled with laughter and jeering. Someone in the group shouted, "You should talk! Do you even know your supervisor's name?" Then, turning to the group he said, "How about Barry? He does even less." The laughter continued and eventually tapered off.

"How about John?" someone shouted to the back of the room. As a member of the maintenance crew, John Lowery was not involved in this meeting, which was for operational staff. He was in the canteen before the start of his shift. The men laughed because they all knew it would be the last thing John would be comfortable with, given his quiet and shy nature.

George looked over to John, who was sitting near the wall looking down into his coffee cup. John had been outspoken about his disagreement over enterprise bargaining and allowing the company to change the way it ran its operations at Longford. He had often objected to the company's change to its philosophy behind manning, and the organisation structure. These changes had resulted in operators assuming greater responsibility for the operations of the plant and relieving the supervisors of their leading-hand role.

He would often shake his head and say, "My sister died as a result of a fire, my brother died in a house fire, and the way things are going, I'll die in a fire," as if foretelling his fate. On these occasions, the men had remained silent. They had known of the emotional and physical scars John carried as a result of his little sister being burned and the subsequent operations John had undergone to provide skin grafts to his sister.

John kept quiet and ignored the folly about him. It was as if he had resigned himself to the fact that no one would heed his forebodings, and no one would take him seriously.

Sensing John's discomfort, George spoke up. "Now leave John out of it," he said. The friendly, yet serious admonition from George diverted everyone's attention away from John, as he had intended.

As the noise died down, Terry turned to George and asked, "George, would you like to take on the job?"

The workers applauded and cheered in support — partly because they genuinely liked George and partly because it would take the pressure off them and put the matter to rest.

George smiled and rubbed his chin before answering. "Well, I'm not exactly what you'd call a union man, Terry, actually, far from it. Before joining Esso, I was a market gardener and small business owner, not exactly breeding ground for the unions."

Terry pleaded a deal with George. "Just do it for twelve months, George. Then we'll have another election. Don't worry, you won't be stuck with the job for too long."

George continued to rub his chin while thinking through the offer. "Righteo, I'll do it for twelve months," he said, much to the relief of the rest of the men. They cheered and applauded George as they steadily stood up and walked out of the canteen to start their shift.

Terry shook George's hand and thanked him. As the two men stood at the front of the lunchroom, planning the next meeting of the union at the local town hall, Barry walked past and shook George's hand, "Good on you, George. You'll make a great shop steward," followed by Charlie, who said, "George, I'm buying lotto today. You've changed your colours."

George smiled and replied, "Yeah, well stranger things have happened."

CHAPTER THREE

BILL SHORTEN IS ELECTED UNION SECRETARY

"STRONG CONVICTIONS PRECEDE GREAT ACTIONS"
James Freeman Clarke

On a cold and rainy day in July, in West Melbourne, Bill Shorten leaned his shoulder against the wall of his small, time-worn office of the Australian Workers' Union as he looked out onto Spencer Street. Even the bleak weather couldn't dampen his spirit. He had just received news of his win in being elected, unopposed, for the position of the Victorian Secretary of the Australian Workers' Union. He was pleased. He was very pleased. It was a big thing for him personally and for the union delegates who had encouraged him to run for this position.

As he relished his victory in the silence before family, friends and colleagues started to call or stop by to congratulate him, he thought about the past few years since graduating from university. While he had enjoyed his job as a solicitor, his passion was for representing workers as a group. He had become sick of asking injured workers to pay for legal cases. He wanted to work out how he could prevent people from having to engage lawyers to begin with; how he could create circumstances where people didn't get injured at work or find themselves unfairly dismissed.

He liked the law, but it wasn't as satisfying to him as being a union organiser. As a volunteer with two of Australia's most venerable unions, the Iron Workers' Union and the Australian Workers' Union, he concluded that he had found his purpose: to serve workers through the unions.

He smiled as he remembered that only four years ago he had joined the Australian Workers' Union as a trainee organiser under the Australian Council of Trade Unions Organising Works program.

Now, at the age of thirty-one, he was the leader of the State division.

Bill had lots of ideas about reforming and modernising the AWU, lifting its membership and raising the confidence of the members in the union. He believed that these were the things that would make the AWU strong.

Bill smiled as he thought to himself, *well, I'm certainly carrying the baton for my family.* He was thinking of the people throughout his life who had been active in the movement. His mother, who was a teacher and worked in universities; his father, who was with the Australian Institute of Marine and Power Engineers and associated with the Federated Painters & Dockers Union; his grandfather, who had been in the printers' union; and his grandmother's cousin, who was secretary of the Seamen's Union for twenty years in Victoria.

Then he thought of the notion of social justice, which had been ingrained in him throughout his school life by the Jesuits within the Catholic schools. Bill had concluded that even if he did not agree with everything the Church taught, he could not argue against its ideal that every person has something special in them and deserves the opportunity for a fair go.

Bill's peaceful contemplation was interrupted by a heavy and loud knock on the office door. Co-workers and colleagues streamed in to congratulate him on his win; some slapping him on the back, others shaking his hand and hugging him.

As the crowd moved away and Bill was left with the union's support staff, his first request was for a meeting of the delegates.

The mood during the delegates' meeting at the Victorian Trades Hall was upbeat. The delegates applauded as Bill walked onto the podium to give his first speech since his election. Bill knew he had made the right decision to leave state politics and continue to work with these people who had befriended him, relied on him and taught him many things over the past four years. He looked down from the podium and saw George Parker standing in the crowd and applauding. Bill remembered him as the reluctant delegate from Esso at Longford, who was doing a great job.

George smiled and lowered his head slightly to greet the new secretary. Bill smiled back. He considered George to be the ideal shop steward: one of the best workers in the plant, honest and respected — a moral compass. He briefly thought of the words that summarised George: integrity, competence and strength.

In response to the applause, Bill looked at the crowd of delegates, raised both hands in a winner's stance and then brought his hands together as a sign of the gratitude he felt towards the delegates. As the applause quietened down, Bill started his speech by thanking the delegates and inviting them to sit down.

"First of all, I want to say thank you for your support and belief in me. This is not my win. This is our win. I feel, privileged and humbled by the support you have given me. Our union, the Australian Workers' Union, has a proud history as one of the oldest unions in Australia. A union with a record of representing people from diverse backgrounds, professions and locations throughout Australia. A union with its origins rooted within Australian labour history dating back to the 1880s. Unfortunately, in recent years a lot of the members feel that they have been let down by, perhaps, officials in the past. They are hardened as a result and they want to see more from the union."

Bill continued. "The union amalgamations in recent years have been pretty political. Some have said that the unions have taken their eyes off the ball of conditions for members. Now, I don't think that's fair. But let's face it, we've had a few scallywags come through the ranks over the years." Some delegates nodded thoughtfully, some smirked at the word scallywags and others looked at him attentively.

As he surveyed the room, he knew the stakes were high. He would either redeem the union from some of its past failures or it would decline further. He wanted to make a positive and lasting impact on the union and on the delegates.

He continued in a deliberate and controlled voice. "But I want you, the delegates of the union and the members, to have pride again!" The delegates applauded, and Bill waited for the applause to subside before continuing. "We can only do that through focusing on the industrial needs of our members, and politics becomes very much a second order issue."

Some of the delegates shouted "Yes!" in agreement.

Bill knew the art and value of starting with the end in mind, so he painted a vision for the delegates as he continued to address them. He said, "My aspiration is that in ten years' time, we will convene a meeting of this very group and that you will say to one another and to me, 'Do you remember when we campaigned on this wage outcome or that safety issue?' That's when we will know that we have been successful in rebuilding the union; we have reclaimed our pride in who we are and what we represent, and the union is stronger!" The final comment roused the delegates to a standing ovation. Bill Shorten had managed to restore their faith.

While Bill felt excited and privileged, he also felt daunted by the task ahead. To break down the resistance of the once faithful and to rebuild confidence in those who felt disillusioned by what they had believed in was no small feat. But he knew from his past victories that these feelings would propel him to overcome any obstacle in the way of triumph, especially when it was for the good of others.

On the drive back to the office, Bill remembered one of his first tasks as a union organiser, it was to organise a scrap metal foundry where there had been a death and many severe burns and injuries. When Bill arrived at the foundry in Melbourne's north, he met with the workers and sensed a reign of fear and intimidation. Most of them were old and had come to Australia by boat. They spoke little English. He realised that they were reluctant to talk to him not because of a lack of interest, but because they were terrified of the consequences after he left. Bill's determination was strengthened by his hatred for such a workplace regime and his desire for fairness and equity. He came back to the foundry five times and eventually unionised the entire workforce. Then, he started a campaign to improve the safety and working conditions of the workers.

This victory, along with so many others similar to it, increased Bill's self-confidence many-fold and he knew that as long as his motives were right, he would be successful in his new role as secretary.

CHAPTER FOUR

RONNIE RAWSON

"HAPPY IS THE MAN WHO FINDS A TRUE FRIEND, AND FAR HAPPIER IS HE WHO FINDS THAT TRUE FRIEND IN HIS WIFE."

Franz Schubert

Ronnie Rawson was fifty-two years old and content with his life, especially with his family. He was born in Melbourne and lived there with his grandparents until he was fifteen years old. He then moved to Maffra to help his parents run their dairy farm. There, he met and married one of the local girls, Rhonda. He loved her now as much, if not more, than he loved her then.

After high school, Ronnie commenced working for the Department of Supply at Maribyrnong and over the years held a number of jobs, including a maintenance position with Maffra Engineering performing maintenance at the Esso Longford Plant as a contractor. After a few other ventures, Ronnie started working for Esso as a trainee operator and progressed to the position of Operations Technician Level 2.

Initially, the working hours and rosters weren't that great. For ten years, he worked on the three-shift roster, rotating days, afternoons and nights with a few weekends at home with his family. Consequently, Ronnie and Rhonda's friendships were impacted. They lost contact with some friends they would have liked to have kept in touch with. However, the close ones remained and persevered with Ronnie's working hours.

Ten years after Ronnie started working for Esso, the company changed to the twelve-hour shift roster. Ronnie thought it was terrific. The quality of life changed for him and his family. The new shift arrangements meant that he could enjoy weekends off with his family and friends.

He thought his work was interesting despite the mental demand that comes with dealing with numerous volatile products. He enjoyed the variety of work, change of areas and the multi-tasking. His work included surveillance of all sorts of machinery which ran constantly — pumps, heaters, turbines, engines and so on. While the equipment ran twenty-four hours a day, it needed to be monitored for breakdowns or seal failures. Surveying and checking the machines for correct temperatures, pressures and strange noises or leaks and submitting work orders for maintenance provided the variety of work Ronnie enjoyed.

What he enjoyed most was the genuine mateship between the workers on the plant. He would often say to family and friends, "The greatest thing about working for Esso is the extremely friendly environment and mateship. We're a close-knit group."

Some of the friendships at work extended to occasional social outings outside of work. Because the work times and days off of the workers coincided, it wasn't a rare occasion for some of them to organise family outings, fishing or four-wheel drive trips, a hike or a couple of days in the snow. These occasions were very dear to Ronnie and Rhonda.

Ronnie's gentle and kind demeanour made him a likeable person. Like the rest of the workers, he had good relations with management. Most of the supervisors were operators who were put in charge of a group. No one came from an office environment. The workers knew their background and they knew the roles and jobs of the workers. That was the natural progression for those who were capable and interested in progressing in their careers. That system worked well and the workers felt comfortable with those arrangements because they knew that the supervisors were knowledgeable.

Over the years, though, things changed. When enterprise bargaining came in, management clamped down on many things. They reduced the number of supervisors on each shift. They changed the maintenance system and, as a result, lost a lot of the maintenance personnel. This affected maintenance work and built backlogs. Urgent maintenance got referred to the morning meetings of management, where it was discussed and prioritised.

If there was a situation that affected operating the plant, the operations supervisor was advised and he would decide to by-pass raising it at the morning meeting and get maintenance straightaway or leave it to discuss at the meeting.

One evening, Ronnie and Rhonda were coming back from their daily walk. He had his arm around his wife's shoulder and she had her arm around his waist. This was a sign of their attachment to each other, even after many years of marriage. They talked softly as they meandered along the footpath in the street where their family home was situated in Maffra. They had been together for a long time and knew they could always rely on each other.

Ronnie was reminiscing about life and Rhonda listened, as if for the first time. They walked slowly and enjoyed each other's company as Ronnie spoke.

"You know, love," he said "growing up in Melbourne and living with Grandma and Grandpa was very nice. But I'm glad I moved to Maffra." Rhonda smiled as she looked ahead. "Imagine if I hadn't, I may never have met you." Then, he turned and looked at Rhonda and said, "I wish I could turn back the clock and find you sooner and love you longer."

Rhonda was surprised at her husband's poetic eloquence. She smiled, looked at him and asked, "Did you make that up, Ronnie?"

He laughed "No, I didn't. I'm not sure who said it but it's how I feel." They laughed and held each other tight as they continued their walk.

Rhonda said to Ronnie, "Well, I feel the same way about you, dear. Our life is peaceful. We've done a pretty good job of raising wonderful children we're proud of and it's not long now before they start their own lives and you and me can have some us time."

Ronnie smiled and nodded. "Well, thank God we enjoy each other's company. You're such a gentle and patient wife. I can't wait to spend the rest of my life with you," he said as he drew her closer.

Back in their home, Rhonda sat in the lounge room looking through family photos. As each one brought joyful memories of some part of their past, she would comment on it to Ronnie as he sat by the fireplace reading the evening newspaper. Eventually, as the evening drew to a close, Ronnie folded the newspaper, placed it on the side-table next to his recliner and stood up, saying, "Good night, dear. I've got to start work early tomorrow so I better get to bed."

Rhonda replied, "Oh, good night. I'd forgotten you changed shifts with George. It's a good thing he and Judy are taking that fishing trip with their friends."

"Yeah", Ronnie agreed. "Give me fishing with my beautiful wife and friends any day," he joked.

CHAPTER FIVE

ROBERT MILLER

"GOODNESS IS THE ONLY INVESTMENT THAT NEVER FAILS."

Henry David Thoreau

Rob Miller was a local boy who had lived in the area all his life. He came from a large family of nine. He attended technical school and completed year eleven. After leaving school, he got an apprenticeship as a motor mechanic and worked as a mechanic for six years. That employer went broke and Rob got a job in Maffra for two years, after which he joined Esso as an operations technician. Up to the date of the accident, Rob had been with the company for seventeen years.

Rob and his wife had three children. Twin boys, who in December 1998 were going to turn twenty-five, and a daughter, who was seventeen years old.

Rob got married when he was nineteen years old. That year, they had their twins. The first few years were tough for the Millers. Rob described the initial few years of work as "Just a way of just getting money." However, when he started working for Esso, it was good money. It made such a difference to his family life.

He would often joke and say, "My first year with Esso, I paid more money in taxes than what I'd earned before, so we were thrilled!" The good pay, in addition to the employee benefits, such as the twenty-five percent subsidy on company products, health care and education assistance to help send the kids to university, made all the difference to their quality of life.

However, over the years, most of the benefits stopped. Rob described the first seventeen years with the company as 'good years'. Despite the few strikes and minor disagreements, the workers and Esso management enjoyed a respectful relationship. At times, Rob would think about the difference working for small employers in comparison to working for Esso.

He often thought that even though the pay and conditions were good at Esso, the first two smaller companies had other good aspects. For example, the person that owned the motor mechanic business used to come to work each day and knew Rob personally. He knew who the good workers were, and everyone had some give and take. The workers would do things they weren't required to do, and the owner would give the workers paid time off even though he didn't have to. The second company Rob worked for was a bigger organisation but the bosses were still a bit more flexible and there too was that give and take between management and workers. But at Esso, which is a 'great big thing', everything was by the book. This made the workers do the same.

All in all, though, Rob Miller was quite satisfied working for Esso.

CHAPTER SIX

PETER WILSON

"'CAUSE WE WERE JUST KIDS WHEN WE FELL IN LOVE"

Ed Sheeran , *Perfect* Asylum. Atlantic

Peter Wilson was the life of every gathering. His wife, Locky, of almost thirty years and their three sons, James, Brett and Luc, were his reason for living. He had centred his entire life on them and their interests, school and sports.

He would often say to Locky, "Darlin', if you can't be with them, I have to be. Our boys will always have one of us with them. It's the only way to care for them."

Locky, the serious one in the marriage, was matter-of-fact. She would nod and say, "Yep. That's how we've done it for years and that's how we're going to do it for the rest of our lives."

Love for Peter and Locky had started when he was thirteen years old and she was eleven. They often reminisced about the day Peter and his family moved in to the house across the road from Locky's family home. Their childhood friendship had blossomed to teenage romance and eventually to marriage in 1969.

Locky would tell stories about their life to the boys, especially when they were growing up. "Boys, if you ever want to know the best memory I have of your dad, all I can say is it's my whole life. The only thing I can say is that we grew up together. We made our deb together. He got his license and then I got mine. We went to the movies, the dances and all sorts of places together. Everything we did was together. Our life has been intertwined with one another. I can't wait 'till you boys are grown and your father retires. That's when we'll go travelling together and have a rest."

Peter would reach out and put his arm around Locky's waist. "Yup, she's the best friend I've ever had."

The boys would roll their eyes and keep eating their meal, not realising that this family bond was not a common experience for a lot of other children.

Brett, vying for attention, would at times interject by asking, "Yeah, but how many times have you packed your bags Mum?"

Both Peter and Locky would look at each other and Peter would say, "Twice mate. And that's not a bad inning for almost thirty years of marriage. Not bad at all. I reckon she's a keeper."

Locky would slap Peter's shoulder and say, "Oh shut up, Pete! You're lucky I'm keeping you!"

Wit and humour characterised the Wilson family, along with unconditional love, security and solid resilience.

Peter was from a small, cultured and refined Protestant family which had a love for fine music. His mother was an opera singer and his father was an appreciator of the opera and classical music. This ingrained a love for instrumental music in Peter and, in turn, he encouraged the boys to play musical instruments. He would take them to all their music lessons and often to concerts in Melbourne.

Locky came from a large Catholic family, who were fun-loving and caring.

Peter had an uncanny gift for fun and friendly teasing. James would often tell him, "Dad, you're such a shit stirrer."

Peter would laugh and ask, "Is there any other way to be?"

There was rarely a time in the Wilson household when Peter didn't give cause for laughter.

Sometimes, he was the brunt of the jokes and teasing by the boys. They would either tease him about his height or baldness. At five feet ten inches or as Peter would say "178 centimetres" to sound taller, he was shorter than the boys, who averaged six feet two inches. He would stare at them with his piercing blue eyes and nod slowly. "I reckon I'm gonna get my own back when you three little shits lose your hair. I'm gonna take a photo of each of you and make it into posters to put around the house."

They would all break into laughter and one of them would retort. "Yeah, but we won't be carrying your belly," referring to Peter's rounded stomach.

Locky would shake her head and say, "You boys are bloody rude to your father. Let's start dinner before I refuse to feed you."

Peter and Locky had a typical marriage, characteristic of the period they had been married in. She did all the housework and he carried on with the 'man's work' outside the house and on the sheep and cattle farm.

No matter how tiring farm work was, Peter drew so much joy from it that he always found time and energy to do it.

At one stage, they were farming cows. Peter would have to dehorn them to stop them from attacking people and other cows. On one occasion, he called Brett over and asked, "Brett, son, do you know much about dehorning?"

Brett scratched his head, dreading the invitation to help that would be forthcoming. "Nah, not really Dad," he replied.

Peter explained that the process isn't comfortable for the cow, nor the farmer, but it had to be done. Then, while locking the cow in a restraint to minimise the struggle, he said, "Sometimes, this blood spurts in a fine mist spray that you can't see."

Once the cow was secured and he had removed the horn, he winked at Brett and called Locky, on the pretence that he needed help. Brett wondered what his father was up to.

When Locky came over, she walked straight into the invisible gushing mist of blood. Peter let out a roar of laughter when Locky jumped as the blood hit her face. She stood still for a few seconds and wiped her face. Brett dreaded the fallout this would cause. Locky looked at Peter and bellowed her favourite reprimand, "Piss off, Peter! That's such a stupid bloody thing to do!"

Peter kept laughing as if there was no other way to react. He shook his head and kept working while Locky walked back into the house, muttering a barrage of swear words.

On another occasion, Peter was running a farm of 250 acres with sheep on it. At one stage he had 700 sheep. Most of the family weekends were spent crutching, cropping or shearing the sheep. Peter would do most of the work, with the three boys helping; James being the older one and more willing than Brett and Luc.

One year, he had contracted some shearers to shear the sheep. After paying them, he had said to the family, "You know what, next year I'm going to have a crack at this." When the time came to shear the sheep the following year, he stayed true to his word.

First, he had to learn how to shear sheep. He decided to follow step-by-step instructions in a book as he sheared his first sheep. He pulled out the first sheep and secured it between his legs. He read each stage of the instruction book and followed it step by step. After shearing the sheep, he released it down the chute. Then he had to skirt the fleece. He reached under the table to collect the fleece, but it fell apart. Peter stood there looking at the broken fleece, wondering what had happened. He muttered to himself, "What the heck? I read the instructions step by step. I wonder what's wrong with the fleece."

Then, checking the instructions, he noticed the sequence of the pages, one, two, six, seven and so on. Roaring with laughter, he realised the instructions had most of the steps missing.

Peter had worked for Esso for twenty-nine years. Over the years, he had gradually climbed the ranks to the level of Maintenance Superintendent. In this role, he was responsible for several functions, including the Planners, Clerks, and Services & Waste Management, Contracts Administration, Technicians, Instrumental/Electrical Supervisors and Mechanical Supervisors. Peter was known at the plant for his fun-loving personality, loyalty to Esso and his ability to take new employees under his wing and mentor them. He was highly respected as a friend, mentor and leader on the plant. Most of all, he was loved like a father or older and wiser brother by most.

CHAPTER SEVEN

JOHN LOWERY THE BROTHER

> "I HAD A BROTHER WHO WAS MY SAVIOUR, MADE MY CHILDHOOD BEARABLE."
> *Maurice Sendak*

John Lowery had aged beyond his years. The layers of unhealed trauma since childhood had taken their toll. Yet, beneath the pain and sorrow, lay kindness, compassion, wit and a subtle intellect. He was a man of habit and routine. Every day started with an early cup of coffee in the small kitchen of the unit he had shared with his mother before her recent death and now shared with his youngest sister, Kerry. Then, he would shower and dress before starting his short walk to the bus stop where the Esso bus would meet workers going to Longford.

After work, he would get on the same bus to take him back to town. The Gippy, as it was affectionately known by locals, was his first stop. There, he would greet the men he had made good friends with over the years and, after a few beers, a friendly dart match or a game of pool, he would go home.

This routine contrasted with two of his passions. The first was his involvement with the Country Fire Authority, which had district headquarters in Sale. He was heavily involved in the marches, parades and the competitions with other country authorities, helping others and training new volunteers. In March 1998, John received recognition for his service to the CFA. His second passion was observed when his beloved football club, St Kilda — affectionately referred to as "the Saints" — was playing. Most of the time, John was in control of his emotions; the only time he strayed from this self-control was at a Saints' game. Win or lose.

One afternoon after work, he walked into the unit and saw Kerry sitting at the kitchen table. As he greeted her, she looked up at him slowly. She looked frightened. He had noticed this look gradually form part of her appearance since their mother's recent death. He saw a deep, haunting sadness in her eyes.

"Hi, Bubby," John said, calling Kerry by the pet name he had given her at the time of her birth when he was twelve years old. "How are you?"

Kerry was slow to answer. John noticed a quiver in her voice. "I'm OK, John," she replied.

"What's wrong, sis?" he asked, less jovial than when he came in.

"Ah, nothing's the matter, dear," Kerry replied. "I'm just thinking of ... I'm thinking of going away for a while. I'm going to Perth ... you know, for a change of scenery." Kerry's frail laugh betrayed her effort at hiding her grief and torment.

John sighed, deep and loud. "Is this about Mum's death?" he asked.

Kerry cupped her face with her hands and started to sob uncontrollably. Unable to speak, she whimpered.

John waited patiently for her reply. "You right? What's going on?" he asked.

Kerry took a deep breath and replied. "It's all in my head, John, it's all in my head," she said slowly, as if tormented by a darkness of the past, "the memories of our childhood, the abuse, the shouting and the fighting. Mum and Dad drunk and yelling at me and Bruce. Annie trying to protect us. Me and Bruce walking in the dark as children, frightened to death as we walked to your house for safety. I just can't get these things out of my head. And now with Mum gone, I just feel so lost and ... I don't know ... just lost." She sobbed more forcefully and was unable to speak.

John was speechless at first. He was very familiar with the stories and scenes Kerry was describing because he had lived the past she was talking about.

"What's bloody Perth gonna do for you? How's a trip to bloody Perth gonna fix those things in your head?" he asked.

Kerry slowly and gradually stopped sobbing and carefully considered her answer. The truth would be unbearable for a family united in so much trauma, grief and love. She took a deep breath, and with the same quiver in her voice as when she had started the conversation, she replied, "The doctor says it's going to be good for me to take a break."

Hiding the truth from her loved ones was the hardest thing Kerry had been faced with in recent years. Her extreme depression had led to her anorexia nervosa and the doctor's attempts to treat her with medication and therapy had failed miserably, leaving her suicidal. She wasn't sure if she was keeping this a secret from everyone to shield herself from the shame small-town living can bring, or to protect her family from the horrors of the depression she lived with. Mental illness can be so lonely and isolating.

In the ensuing silence that lingered between the siblings, Kerry had flashbacks of the final conversation with her treating doctor, who had said, "Kerry, there is one more treatment I would like to recommend for you." He had hesitated, knowing that past discussions about treatment had started with hope but had resulted in dire despair and deeper depression. Kerry had not responded. She had looked at the doctor with a blank stare in her beautiful blue eyes. She was exhausted from the burden of the depression and not eating. "I'd like to recommend Electric Convulsive Therapy," the doctor had said slowly and calmly, to compensate for any panic these words might conjure up in her.

Kerry had suddenly realised he was talking about electric shock treatment. She was silent for a short while longer. Then, she looked away from him to the multi-coloured wall chart of the human anatomy hanging on the wall behind the doctor's chair.

She had asked in a frightened and weak voice, "Is that electric shock treatment?"

The doctor had looked down at the silver pen he was holding and rolling with his thumbs and forefingers. He had sighed deeply and replied, "Well, yes. It is known in the community at large as electric shock treatment." He had continued, "It's relatively painless because these days we anesthetise patients before we perform the procedure. It has proven, in many cases, to be a fast and effective treatment strategy for mood disorders."

"So how long you going for?" John asked, interrupting Kerry's thoughts. By now he was sitting opposite her, leaning forward as if trying to reach for her pain and take it away as he had done so many times before throughout her childhood.

"Well, I'm not sure. As long as it takes ... I suppose," she replied. "John, I want you to promise me that you'll look after yourself," she continued slowly. "I love you very much. You and Annie were like parents to me and Bruce when we were growing up and when Mum and Dad were off their faces drunk and not coping with life." She started to sob again.

John sighed deeply and slowly sat back in his chair. His instinct betrayed the truth. Kerry was not well. Really not well. He hoped the trip to Perth was not the prelude to her harming herself or, God forbid, taking her own life.

As his sister sat there crying with her hands cupping her face, he felt helpless. He wondered what he could do for her. Then, in despair he said, "I'm gonna lie down for a bit." John had learned the art of escaping pain through sleep from an early age. He stood up slowly, and quietly walked to the old, brocaded recliner chair in the lounge room. The chair he had bought his ailing mother so she could rest while she looked out of the window onto the street. He sat down and stared out of the window. He pondered how children can stay loyal to and love parents who have hurt and abused them throughout their life. The thoughts lingered for a while and gradually faded as John leaned back in the recliner, resting his head and back.

- - - - - -

Kerry and Bruce were knocking at the door as fiercely as two children in a state of panic and terror can. John jumped off the chair he was sitting on at the kitchen table and opened the door. His poor baby sister and brother were accompanied by the older sister Annie, who was eight years younger than John. The younger children were dressed in shabby pyjamas and shivering from the cold. Their wide, frightened eyes, unwashed and unkempt hair betrayed another day of alcohol bingeing by their parents, undoubtedly resulting in violence and terror towards the children.

Annie was crying and Kerry and Bruce were cuddling one another, shaking with terror and cold. John let them in, wrapped each one in a woollen blanket and asked what had happened. Annie described what had happened on this occasion. "I came home to check up on Kerry and Bruce like I always do. I found them home alone the poor things; Mum and Dad weren't home. The house was freezing cold and there was no dinner cooked. There was fish out and I didn't know how to cook it, so I made some cold meat and salad for the kids. Just then, Mum and Dad got home. They saw that I hadn't cooked any dinner and they started yelling and throwing things around the house. They yelled at me for not cooking the fish. Dad said he would break every bone in my body. Then he looked like he was going for Bruce. I was so scared.

"So, I grabbed Kerry's and Bruce's hands and we ran out of the house. Dad chased us all the way down the street. We kept running. Then, the yelling stopped, and he started muttering as he turned and went back to the house. We didn't know what to do. We were cold and scared. We were too frightened to go back in the house. The yelling in the house started and Mum and Dad were fighting again, we could hear them from far away, glasses smashing and everything." John looked into Kerry's eyes and wondered what this daily terror was doing to these innocent children.

The scene gradually faded and slowly John was awakened out of his slumber by gentle tapping on his shoulder. At first, it was hard to open his eyes. He blinked continuously and rapidly until he slowly woke up. As he opened his eyes, he saw Kerry looking down at him with the same frightened and meek eyes he had just dreamed about. He looked around the room for a short while, uncertain of his surrounds and wondering what was happening. Then he realised he had fallen asleep and had been dreaming. Since childhood, John's dreams were not imaginings. They were the horrendous scenes of his past, a daily reminder of the fear and wretchedness that had been the legacy of being a Lowery child.

Still dopey from the deep sleep, in a husky voice he asked, "What's the matter, Bubby?"

Kerry spoke gently and quietly. "Annie just called. They want us to go over for a family dinner. She's made roast lamb, veggies and dessert." Kerry was comforted by the invitation to join her lifelong protectors and consolers for dinner, but she did not feel she had the psychological strength to sit through a family dinner, despite her love for John, and Annie's kind and loving family, Rob and Mattie.

"Oh, OK. I'll change, and we can go," John said as he straightened himself in the recliner.

Kerry hesitated. And with trepidation, she said, "John, I told Annie I won't come. I haven't been feeling the best and I'd rather have a quiet one tonight."

John looked into his sister's beautiful eyes. They were a window to her grief. His enthusiasm to go to Annie's was dampened. He wanted to encourage his sister to reconsider her decision not to come but decided to let her be. He nodded and said, "OK, Bubby. You stay home and rest. I won't be out too late."

Kerry nodded with relief and said, "I'll drive you over. Annie said she'll bring you back home."

John sighed, stood up, and said in a soft voice, "OK, sis."

CHAPTER EIGHT

JOHN LOWERY THE IN-LAW

> "THEY SEEMED SO UNITED THAT
> I LOVED THEM AS ONE PERSON."
>
> *Robert E. Lee*

Brother and sister strolled to the car. As John got into Kerry's car, he smiled and asked, "Hey Bubby, do you remember the first car you bought?"

Despite her depression, Kerry found reason to smile. She nodded. "Yep, I sure do," she replied with a chuckle.

John continued. "You never told me why you didn't bloody ask me to help you find a decent car," he teased.

Kerry smiled and said, "Well, I was trying to be independent, adult and all. I mean, why else would a naïve country girl buy a car without asking her mechanic brother for help?"

They both chuckled.

Kerry remembered the event very clearly. John was a mechanic by trade. In her effort to gain independence, she had gone to a second-hand car dealer and bought an old-fashioned Mini Minor. To start it, the ignition key needed to be turned while simultaneously pushing a button on the floor with her foot. Proud of her purchase, she had driven the car to John's house. He had looked at it and snarled. "You bought a piece of junk! You should have asked me to help you." She had defended her decision and told John what a bargain she had picked up. Unimpressed, he had kept working on his own car in his home garage.

A few days later, Kerry had noticed a hole under the pedals of the car. Coyly, she had taken the car to John's place, so he could fix the hole. She had known that despite his disapproval of her buying the car without asking for his advice, he would help her. She was right. As soon as she had told him about the hole, he had started working on it. He had drilled a plate over the hole to stop her foot going through, "like the Flintstone's car" as John had quipped. After a short while he had emerged out of the garage and said to Kerry, "Take it for a test drive, sis, should be right now."

After driving for a short distance, Kerry had realised that the brakes were not working. Despite her panic, she had stopped the car and slowly driven back to John's place, "there's no bloody brakes," she had said.

John had frowned at her and replied, "Yes there are. Let me have a look." After a quick look, he realised that he had drilled a hole in the brakes when he had drilled the plate in. So, he fixed the brakes too.

"Do you remember the brakes, Bubby, and how I put a hole in them?" he asked as he laughed and shook his head.

Kerry nodded. "Yeah ... I do." She responded as if the nostalgia had brought some reprieve to her living nightmare of depression.

Rob was waiting at the front door when Kerry and John drove onto the driveway. He had been such a pillar of strength to the family from the day Annie had met him and introduced him to them.

As Kerry slowly drove up the driveway and stopped the car, Rob approached her. Kerry wound down the window and greeted Rob. In turn, he looked at Kerry with the usual warm and friendly smile and asked, "Are you sure you won't come in?"

Kerry nodded. "Yeah, I'm sure, Rob."

After the routine farewells, Kerry reversed the car onto the street and drove away, and John and Rob walked into the house.

Dinner was always fun at Rob and Annie's, but this night was an exception. After dinner, Rob cleared the dinner table while John washed the plates and cutlery, his custom at all family gatherings. Annie prepared the apple pie she had baked that afternoon and served it with ice cream. To everyone's surprise, Matthew said he had homework and excused himself from dessert.

Having finished the washing up, John went back to the dinner table and sat down as his sister served dessert; first to Rob, then to John and finally, herself.

John methodically ate his dessert, first by slicing a small piece of the apple pie then by scraping some ice cream and combining these on his dessert spoon. While the silence between the three adults was peaceful, Annie sensed some unease in John.

"What's wrong, John?" she asked.

John stayed silent as he considered Annie's question. Then he looked up and smiled at Annie before replying, "You always know, don't you, Sissy?" he asked.

Rob remained silent. He knew that some conversations between siblings were sacred.

Annie kept eating her dessert slowly as she patiently waited for John to continue.

"When you called Kerry, I was napping. I had another one of those dreams," he said. Then after some silence, he continued, "The story was from a different time in my life but the same stuff; the kids were young, and Mum and Dad were drunk. I'm so tired of it all. And I don't know if Kerry's told you, but she's going to Perth. I bet you it's to get away from it all. I don't blame her. I wish I could do something to bury the past and see a happy life in my future," he said, his voice trembling.

Annie nodded as she looked down at her plate, quietly scraping the remaining dessert to one side. She placed her spoon down on her plate and looked up at John.

He continued. "I mean, how much tragedy and sadness have we had, right back on Dad's side of the family? Dad being abandoned and being brought up by another family, Gaye's death, Bill's death, Dad dying and now Mum."

Annie and Rob knew these events of the family's history and the deep pain that they had etched into the lives of each child.

During the quiet that followed, each person around the dinner table silently reflected on the same events as if in unison.

Rob recalled being told about the single event that triggered the instant decline of both Mabel's, more affectionately known as Win, and Frank Lowery's mental health. The incident occurred at a time when there was no social support for people in such circumstances. In turn, this incident had disabled their ability to continue to function as healthy human beings and as the nurturing and loving parents they had been up to that point. Instead, they had numbed their pain by turning to alcohol for comfort, which invariably resulted in neglect of, abuse and violence towards the children.

He could almost hear Kerry telling him the story.

"Rob, do you know about our sister, Gaye?" she had asked. Rob shook his head in response. "Well, I didn't know until I was eight years old, when Mum told me. I was the youngest of seven children. We lived on a farm in Maffra. When I was just born, one of my sisters, Gaye, and my younger brother Bruce, were outside playing. Gaye was about four years old, and Bruce was two years younger than her. Mum was inside looking after me. Apparently, Bruce had run inside and said, "Gaddie on spider." Mum had thought there was a spider on Gaye. She had looked out the kitchen window and she had seen that Gaye was on fire.

"As it turned out, she'd had found an old-fashioned lighter and flicked it. The lighter had sparked and because Gaye was wearing a nylon nightie, it had caught fire and just melted onto Gaye's tiny body. I remember Mum telling me that she'd wrapped her in a blanket and carried her across three farm paddocks to the nearest neighbour to get hold of Dad. We only had one car and we were too poor to call an ambulance. So, Dad came home, and Mum and Dad took Gaye to the Maffra hospital. They couldn't do much for her because she was so badly burned, so they drove her to the hospital in Sale. They were told to take her to Melbourne.

"So, my Uncle Barney had driven the car while Dad nursed Gaye all the way to Melbourne — in those days it was a six-hour trip. So, they drove her to the Royal Children's Hospital. They were told that she had suffered third degree burns, where the skin and underlying tissue had been destroyed and wouldn't heal without a skin graft. This was 1961, when skin grafts were a fairly new medical procedure in Australia. Mum and Dad were told that she had to have extensive skin grafting because she was burned on every part of her body except her hands, face and ankles. John was twelve years old and for some reason, he was chosen to be the skin donor. He had to give skin from both his legs, the front and back. He was in hospital for three months. Afterwards, he stayed with my Aunty Mary who took him to the beach every day to try and heal his wounds. In the meantime, little Gaye was in confinement in a room on her own. They wouldn't let Mum see her because they said it would be too emotional for Gaye. Dad and my aunty, who lived in Melbourne, were allowed to see her but not Mum."

Rob sighed deeply as he remembered the end to this tragic story. He recalled that Kerry had stopped at this point to regain the strength to finish the story. Then, she had looked up at him and said, "She died on Christmas Day."

After a deep sigh and with tears streaming down her beautiful face, she had continued, "You wouldn't believe it, it got worse. I remember Mum telling me she had police knocking on our door asking her if she'd killed Gaye because she had too many children. As a parent, she had to deal with the loss of a child, six more to look after and then being questioned about the death of her child." The tears kept flowing as if a tap had been left on. "No wonder life was so hard for them. Mum and Dad weren't supported after Gaye's death. They struggled, so they turned to alcohol and that's when the abuse started. As an adult, I can forgive them for a lot of the things they did to us. They didn't have the help then that we have nowadays." Kerry had stopped talking and had looked away from Rob. Then she continued, "The sad thing is, Mum never drank a drop of alcohol until this time."

Then, Rob's thoughts transported him to the second tragedy the family had faced, eleven years after this one. He remembered Annie recounting the death of their brother Bill, who was two years younger than John. He had died from smoke inhalation in a house fire when he was twenty-one years old. Annie had said that the impact of this had felt like an emotional tsunami on the family, shattering their already fragile minds and emotions.

The silence and Rob's thoughts were interrupted when Matthew ran into the kitchen, carrying a box with both hands and excitedly shouting, "Uncle John, you bought me a crystal set! Wow! Thanks! Thanks! Can you help me put it together?"

"Of course, Mattie," John replied.

Matthew continued in excitement, "Now?"

"Yes, now," replied John.

"What's that?" asked Annie.

Rob turned to his wife and smiling he said, "Oh it's just a really simple radio receiver and doesn't need electricity or expensive batteries." Then he winked at Annie. "You can hardly hear it too so that's good for us."

Rob and Annie watched with curiosity and joy as John and Mattie moved away from the kitchen and sat on the carpeted floor of the lounge room and started the meticulous task of assembling the crystal set. John's kindness and gentleness never ceased to amaze Rob. It was as if John had taken all the hurt and pain and transformed it into love and compassion and the willingness to help.

"Okay," John started to patiently explain the mechanics of assembling a crystal set. "The coil of wire is called a tuning coil. Now, the number of turns of wire on the coil control the frequency. If you wanna tune into different stations, you change the number of turns on the coil."

Matthew excitedly studied every piece of the set that John held up and explained. At times, he reached out to take the piece from John before placing it on the floor to pay attention to another part.

This continued as Annie and Rob started to clean and tidy the kitchen and prepare their lunches for the next day.

CHAPTER NINE

JOHN LOWERY, SHANNON AND ASHLEY

"ONE OF THE MOST POWERFUL HANDCLASPS IS THAT OF A NEW GRANDBABY AROUND THE FINGER OF A GRANDFATHER."

Joy Hargrove

Friday nights were very special for Ashley. At the age of three, she felt grown up and had asked her mother for weekly play dates with her grandfather John.

Shannon was John's daughter. Every Friday evening after work, Shannon would pack the nappy bag for her second child, twelve-month-old baby Bianca, along with Ashley's toys and they would drive to John's house. Ashley told her mother every day that she thought her "Grandpa John was the best and she loved him this much," as she spread her arms wide open.

On the way over to his house, Ashley would ask the same questions in the same order. "Mummy tell me about the time I was born and how Grandpa came to visit me."

Shannon would tell the story, trying very carefully not to leave any of the details out lest Ashley correct her.

'Well, I was in the hospital and you had just been born. I was very happy, and I fell in love with you the moment I saw you in the delivery room. That evening, I had just finished feeding you milk and I had finished eating my own dinner. You were wrapped in a soft, white blanket. The hospital had given you a pink beanie, which was on your head. You were in the clear plastic hospital bassinet with your name on it and I was watching TV.

"To my surprise, I saw my dad —"

"My grandpa!" Ashley would interrupt as if claiming him.

Shannon would continue, "My dad and *your* grandpa walked into the hospital room. He said hi to me and asked how I was. Then he walked over to you while you were sleeping in the bassinet. He reached out with one hand and stroked your hand gently and softly. After a while, he smiled and said, 'You're beautiful.' Then he kept smiling and smiling and smiling at you. I don't think I'd ever seen him smile that much — ever, ever, ever."

Ashley loved this story. Excitedly she would kick her feet in unison against the bottom of the child safety seat and clap her hands as she would shout, "I can't wait to see Grandpa! I can't wait to see Grandpa!" Shannon would shake her head and laugh as she looked at her child via the rear-view mirror.

Then, as predictable as Ashley's first question, the other questions would follow one after the other in the same sequence as they did every Friday.

Shannon loved Friday evenings with her daughters and her father. She felt that it was their sacred space and special tradition. John doted on the girls and Shannon felt that she had grown closer to her father because of these visits.

"Why is Grandpa called Dr Who, Mummy?" Ashley would ask.

Shannon would repeat the story every time. "Well, Grandpa doesn't talk to a lot of people. He is very quiet. He just likes to keep to himself. When he makes friendships, they are for life. When there's people around him, he doesn't say much until he hears something in the background that he wants to talk about. So, he pops his head up and says 'Who?' So, everyone calls him Dr Who."

Ashley would squeal with laughter and repeat, "Dr Who, Dr Who," until she would remember her next question, "Mummy?"

"Yes, Ash."

"Why did Grandpa call you Fangs?"

Shannon would roll her eyes and shake her head at the thought of this memory.

"When I was four years old, I had to have my front teeth taken out. Until they grew back, I only had my canine teeth at the front. So,

Dad started calling me Fangs. And that became his pet name for me. Even after my teeth grew, and until I was thirteen or fourteen years old, he called me Fangs."

On one occasion, Shannon was surprised by Ashley's new question. "Is thirteen old, Mummy?"

Shannon laughed and answered. "I suppose it is sweetheart, if you're three."

"Why doesn't Grandpa have any teeth, Mummy?" Shannon found this question the hardest every week. She didn't know how to explain gum disease, malnutrition and childhood dental decay due to poverty and the inability to afford dental care, to a three-year-old.

Shannon would reply, "He's got teeth, Ash," referring to her father's upper dentures.

Eventually they would arrive at John's house. Ashley would run to John and give him her special hug while Shannon settled the baby. The conversation always started with John asking, "How was your week, Shannon?" with keen interest in his daughter's life and family.

Shannon would sit down and describe the events of the week or share a story about the children. John would listen intently while setting up a special area for Ashley's weekly milk and biscuits treat.

"There you are, Princess Ashley. Your milk and bikkies are served." Ashley would smile at him and pick up the first biscuit, dunk it in the milk and eat the soggy part while John sat opposite her and watched his granddaughter with pride and joy.

On a few occasions, Shannon had caught him sneaking chewing gum into Ashley's hand despite Shannon's disapproval. Once she had objected by saying, "Dad, you shouldn't give her chewing gum. The sugar's not good for her teeth and I don't want her getting used to chewing gum."

John had looked sheepishly at Shannon and Ashley. He had noticed Ashley waiting to be defended and in his usual humour he had retorted, "Ah … up your nose with a rubber hose, Fangs." They all laughed.

Shannon shook her head and muttered, "That's so immature, Dad" and continued to laugh with her father and daughter. "Honestly, up your nose with a rubber hose. Who says that?"

John responded "Me!" before he and Ashley laughed some more.

"By the way, Shannon, when are you gonna cook some corned beef for me? I haven't eaten any since Mum died," he asked, looking at his daughter with mischievous eyes.

Shannon laughed and said, "Dad, I hate corned beef. I hate the smell. You have eaten corned beef since Nan passed. I made it for you not long ago, give me a break."

They both laughed, remembering the weekend John had brought corned beef over to Shannon's house and said, "Cook me this corned beef for dinner, will you?"

Shannon had looked at the meat and said, "You know I hate corned beef, Dad. I hate the smell of it, it makes me sick."

He had insisted and said, "Just put it in the pot and go out. By the time you get back it'll be cooked." So, Shannon cooked the corned beef for him and chicken for herself.

"So, how's work Dad?" Shannon asked her father.

He growled. "They're always cutting back on staff, even the bloody maintenance was cut back. I started as a supervisor back in March and the friggin' backlog of maintenance was 115%. It's an accident waiting to happen if you ask me. Don't ask me when. But it's gonna happen."

"Oh well, don't worry, Dad. Hopefully, when you retire, and the kids are older, we can all go on a big holiday somewhere. Maybe even overseas," Shannon said.

"Well, I won't be around to see that," John replied as if predicting his untimely death.

"Don't be stupid. Of course you will. Of course you will," Shannon replied with unease, repeating her affirmation as if to convince herself

Their attention was drawn to Ashley as she reached for another biscuit and accidently knocked the cup containing the milk, spilling it on John's trousers. She squealed and started to cry. Shannon and John both comforted her and told her that Grandpa was OK. John asked Shannon to pour another cup for Ashley while he changed his trousers.

Shannon was preoccupied with wiping the table with a wet cloth and rearranging Ashley's special table with milk and biscuits when John came back into the lounge room. She looked up at her daughter to comfort her and show her how clean the special table was. She noticed Ashley staring at something. Ashley's face was solemn. Shannon turned around and saw John standing behind her, waiting for her to finish wiping the table so he could resume his seat at the milk and bikkies table. He was wearing a pair of board shorts. Shannon could not recall a time when John had worn shorts. She noticed the extent and depth of the scarring to his legs, caused by the removal of his skin for grafts to save his sister's life when he was twelve years old. He had scars all over his legs, except one shin.

John noticed the distress on Shannon's face. He ruffled her hair and said, "Don't worry, Fangs. You get used to it." Then he sat down near Ashley and quickly distracted her with funny stories about characters he had made up to make her laugh.

Whenever Shannon and the children left John's unit, they were jovial. Ashley's customary goodbye to her grandpa included a kiss on each side of his face, a bear hug and a wave, in that order. Sometimes, he would call her back for another hug, and while he hugged her he would sneak another chewing gum into her hand. Shannon would pretend she had not noticed.

This Friday night, Ashley hugged her Grandpa tighter and more times than ever before. Shannon wondered if it was because Ashley had seen the scarring on his legs and felt sympathy for him.

Holding the baby in her arms while waiting for Ashley to join her, Shannon turned to her father and said, "Love you, Dad." He smiled and waved goodbye.

CHAPTER TEN

MARTY AND SUE-ELLEN JACKSON

> "FAMILY IS THE MOST IMPORTANT THING IN THE WORLD."
> *Princess Diana*

Thirty-six-year-old Marty Jackson stood in the middle of the five acres he and his wife of twelve years, Sue-Ellen, had recently bought and built their dream house on. He put his arm around her waist and playfully pulled Sue-Ellen closer to him. In response, Sue-Ellen wrapped her arms around him and snuggled her head against his chest. Both looked around the property. It was a vast area of picturesque, lush green grass. This was their new home. The large shed on one side of the property had served as their make-shift house while their dream home was being built. This was their dream coming true.

Marty looked at Sue-Ellen as she continued to study the land. "Do you remember when I first asked you if you'd mind being a shed-dweller until we built our home, love?" he asked.

Sue-Ellen looked up at Marty, shielding her face from the brightness of the sun with her hand. "Yup! I sure do. And I'm glad I said yes. The kids love the space and open air. And your new job at Esso means you can be home every day, even if you're working the night shifts."

"Yeah. It's a great place to work and it's only ten minutes' drive. I love working with the blokes most of all — Pete, Adam and Jimmy Ward. It's the best, Suz." Marty took a deep breath as if to absorb the contentment he was feeling in his life. Then he said, "I've been there four months and I've felt like one of the boys from day one. There's such mateship."

Sue-Ellen gazed back over the trees and bushes surrounding the land and asked, "Speaking about Pete, how's he going? How's Locky?"

"Yeah, Pete says they're all doing great. He talks about them all the time. Brett and James are living down in Melbourne. Luc's working with one of the contractors for Esso." Then Marty laughed. "Pete wants him to work at Esso, but you know Luc wants to stand on his own feet and be his own man and that."

Marty was quiet for a while. Then he smiled and said, "They're good boys you know, Suz. Coaching them was so much fun. They're competitive and strong but they're so nice ... you know, such a good family."

He continued, "This year's Pete's thirty years at the plant, and funny enough, thirty-year wedding anniversary too."

Sue-Ellen was surprised. The romantic side of her prompted her to put her hands together in front of her face as if in prayer. She smiled and said, "Oh Marty, that's so sweet. I look forward to our thirty-year wedding anniversary."

Marty nodded and said, "Me too, Suz, me too," and continued. "Honestly, you couldn't ask for a better boss. He's like the father at the plant. He jokes, laughs and takes the younger ones under his wing. But when the job's gotta be done, we get on with it. He's a top bloke." Sue-Ellen listened and nodded.

I wonder what Esso will do for his thirtieth work anniversary, Marty pondered.

Suddenly, Marty and Sue-Ellen heard the screaming of their four-year- old daughter Sophie. Startled, they turned in the direction of her screams. They saw her running towards them, her long blonde hair flowing and bopping as her short legs carried her across the field. Wondering what she was running from, they noticed eight-year-old Hayden.

"Mummy! Daddy! Hayden got a dead mouse!" she screamed as she ran from behind the shed, through the open space towards her parents.

Hayden was up to his usual brotherly teasing. He was carrying a dead field mouse by its tail. It dangled as he held it out and ran after his sister towards his parents.

Eventually, Sophie reached the safety of her mother, her little body abruptly hitting Sue-Ellen's legs and holding them tightly. Hayden kept running as fast as he had been, still holding the mouse out in front of him. Soon, Sue-Ellen joined the commotion and started screaming, "Hayden put that dead thing down! Put it down!" Realising that this was one of the times Hayden's sense of joy and innocent fun was going to override his obedience to her, she grabbed Sophie's hand and started to run towards the house. "Marty, stop him! That's disgusting! Stop him!" yelled Sue-Ellen from a distance as she and Sophie ran.

Marty laughed out loud. As Hayden ran past him, Marty reached out and put his hand on Hayden's shoulder to stop him. "Now that's enough," he said to his son. Hayden stopped running; still laughing he lifted his hand — the dead field mouse dangling by its tail. Marty looked at Hayden's face and saw sheer joy and excitement. His heart leaped with love for his son, his family and his life.

"Go on, chuck it away, mate. It's full of germs," he said to Hayden.

Hayden giggled as he watched the dead mouse dangle from his hand. Eventually, he said, "Mummy and Soph were scared Daddy."

Marty laughed. "Yeah, they were, mate." Then smiling and shaking his head, Marty said, "I reckon you've had enough fun for the day. C'mon let's go wash your hands."

Hayden kept laughing as he dropped the dead mouse on the ground and followed his father to the house.

Sue-Ellen and Sophie had settled on the lounge for a cuddle. Sue-Ellen gently played with Sophie's hair while Sophie watched her favourite episode of the Wiggles — 'Wake up Jeff'.

When Marty and Hayden walked into the house, Sue-Ellen looked over to Marty and shook her head slowly. "He's going to wash his hands, right?" she asked Marty. Marty winked at his wife and nodded, and Sophie sang the chorus of her favourite Wiggles song, "Wake up, Jeff, everybody wiggling ... Wake up, Jeff, we really need you!"

That night, Marty tucked the children into bed while Sue-Ellen prepared the lunches in the kitchen for the next day. Marty came to her and kissed her forehead. "Thanks for packing the lunches, Suz. I'm on early shift so I'll leave quietly. What are you gonna do?"

"Hmm ... drop off Hayden at school, go to Sale to pick up Sophie's shoes and some groceries. I might stop by the hairdressers, then the usual."

"Alright. I'll hit the sack then. It's an early start tomorrow," Marty replied, not knowing that tomorrow his life would change forever.

CHAPTER ELEVEN

25 SEPTEMBER 1998
TROUBLE AT THE PLANT

> "LACK OF KNOWLEDGE IS THE SOURCE
> OF ALL PAINS AND SORROWS."
> B.K.S. Iyengar

In the early hours of the morning on 25 September, the night shift crew was preparing to go home as the day shift workers started to arrive at the Esso gas processing facility in Longford.

The facility comprised a network of pipelines through which liquid and gaseous hydrocarbons were extracted at the platforms and transported to onshore processing plants.

It had three gas processing plants known as Gas Plant 1, Gas Plant 2 and Gas Plant 3, and one Crude Stabilisation Plant. These were referred to as GP1, GP2, GP3 and the CSP. The gas plants processed raw gas to natural gas which was sent by pipelines from Longford to domestic and industrial users.

The Crude Stabilisation Plant processed crude oil to stabilised crude oil. This is used to produce gasoline to fuel cars, heating oil to heat buildings, diesel fuel, jet fuel, residual fuel oil to power factories and large ships, and for making electricity and other products such as asphalt.

Each processing plant was a conglomeration of complex equipment designed to process the raw gas. For example, the slugcatcher is a device used to remove the energy of the liquid which accumulates because of the condensation of water and hydrocarbons after the raw gas is piped from offshore, cooled and its pressure is lowered. The liquids are referred to as slugs. The slugcatcher also stores the liquids before processing.

Heat exchangers, such as GP901 and GP902, prepared the gas for absorption in the absorbers. There was a series of heat exchangers connected to other critical equipment, each performing its own function.

Absorbers are towers with valve trays inside. These allow gases to travel up and lean oil to travel down. There were other vessels and equipment, such as flash tanks, reboilers and a temperature control system known as TRC3B, as well as level control valves and so on.

Obviously, these vessels, pipes, valves, heat exchangers and other equipment were interconnected. If there was a problem with one, it had the potential of having a domino effect on the others in the process chain.

Every day, toolbox meetings were held in the two plant control rooms at the commencement of each shift — 7.15am on day shifts and 9.30pm on night shifts. These were conducted by the shift supervisors and attended by the operators and trainees working on the next shift.

These meetings were used to tell operators of any issues they needed to know about for their shift. For instance, any maintenance work planned in their area, the gas sales forecasts for the day and of accidents or process problems which had occurred on the previous shift. A safety message was also included. On day shift, the work permits for the jobs planned for that day were handed out to the appropriate operators.

The toolbox meetings for maintenance personnel were conducted in the workshop at 7.30 am. First jobs for the day were handed out, and safety requirements and any special instructions were given.

Among the many jobs in the plant, there were the roles of control room operators also known as panel operators, and the area operators.

The main job of a panel operator was to supply gas on time, on specification at the right pressure to the end user. That was fairly basic and fairly simple, but it required constant monitoring. In addition, the panel operators were responsible for many other time-consuming and time-critical tasks and duties. For instance, merging the total flows of gas from the three gas plants into one simulated flow to provide that gas. If too much was supplied from one plant and not enough from another, it had the potential of upsetting the total balance and causing problems to occur further down the process chain.

Another critical task was the extensive and time-consuming administration work to comply with the regulatory nature of the oil industry. One such consideration was exercising caution to meet the environmental protection laws. For example, the operators had to be careful about the amount of contaminated water they put back into the water table or the type of emissions from the gas plant.

The company had, rightly, implemented procedures to reduce those emissions and protect the environment. However, the extra equipment, such as extra compressors, turbines and recovery equipment required for this meant more work and equipment for the panel operator to look after.

In addition, the workers also covered for each other at toilet breaks, when taking up lunch orders or if carrying out some other administrative task.

One of the key functions of the control room operators was to set, control and monitor certain processes of the gas plant to ensure they were operating properly. This was done using a mixture of equipment. One such piece was pneumatic equipment installed in 1969 when the plant was first built, and the other was a computerised system referred to as the Bailey system, installed much later.

Both systems were installed with alarm panels which gave visible and audible alarms should process variables move out of a predetermined range. On the original system, the audible alarm could be cancelled by pressing a button but the visible alarm would stay illuminated until the controlled variable returned to the normal operating range and the operator manually reset the alarm. On the Bailey system, the audible alarms could be cancelled by pressing a button. However, unlike the original system, the visual alarm would stay illuminated until the controlled variable returned to normal operating range and it would reset automatically.

An important aspect of operating the plant was the collection and storage of historical information relating to the processes.

This was done by several systems. Firstly, the Bailey system recorded information. This information was taken by another system referred to as the Process Information Data Acquisition System and stored in a compressed form in a computer server at Longford. This computer was connected to Esso's data communications network. In this way, the information was available to personnel both in Melbourne and Longford. Therefore, the Process Information Data Acquisition System enabled the operation of the plant to be monitored remotely to a limited extent as the Bailey system was only partially implemented in Gas Plant 1.

The only other chronological record of process conditions in Gas Plant 1 was provided by the Surveillance Information Database System. This system required operators to walk around the plant following a prescribed route and record information from equipment on a Portable Data Terminal. Recordings were taken twice a shift, that is, four times a day for a set of specified process variables. Records from the Portable Data Terminal were then downloaded into a computer system, which was also linked to the Process Information Data Acquisition System.

In addition to these methods of collecting and recording information were the log books kept by operators and supervisors.

The workload on day shift had always been significant, especially since the onset of enterprise bargaining whereby conditions such as manpower had been negotiated to accommodate pay rises. It was something that never sat comfortably with the workers, but it was something that was undertaken to keep up with the cost of living.

Another issue was the reduction of supervisor numbers that resulted in operators having to deal with the overflow of duties for which the supervisor was responsible. In 1987 when Jim started as a trainee panel operator, there were four supervisors. On the day of the disaster, there was one supervisor.

In reality, though, one person could not perform the work of four. Naturally, some of the responsibilities would overflow into the work of the rest of the crew.

On the night shift before the day shift of 25 September 1998, some significant process upsets had taken place within Gas Plant 1. These would create a flow-on effect to other processes in the plant, including temperatures of other equipment, products, and a build-up of condensation and so on, impeding the usual operations of the plant. But no one knew this.

Early in the morning on Friday 25 September 1998, Jim sat on the workers' bus travelling the road from Sale to the Longford Gas Plant. Jim was a panel operator in Gas Plant 1.

Ronnie, an area operator also travelled by the workers' bus.

Their arrival was followed by the daily arrangements set by the company. The bus arrived at the front gate and Ronnie and the other workers swiped their cards to enter the facility. Then the bus drove the workers directly to the Gas Plant 1 control room.

What followed was the start of the shift routine — the handover, which is followed by a toolbox meeting.

Jim strolled into the control room and bade his hello to the night shift control room operator before getting a rundown on what had happened overnight. He was told about some problems they had experienced overnight with some of the equipment.

The night shift control room operator also informed Jim that during the night they had high levels of liquid in the slugcatchers and high gas demands due to the low ambient temperature overnight. However, he said that the liquid flow into the slugcatchers was back to normal and that Gas Plant 1 was now operating normally.

On the same handover, the outgoing area operator told Ronnie of issues with the temperature recorder and controller referred to as the TRCB3 bypass valve and how the two night-shift supervisors had tussled about leaving it open or closed.

During the toolbox meeting of the day shift for 25 September, nothing of any importance was raised about the issues of the night before nor any other significant issues concerning the operation of the plant in Gas Plant 1.

After the handover, Jim assumed control of the Gas Plant 1 control panel. He checked the process conditions and concluded there was nothing he had to change. Everything appeared to be running properly. Then he checked the status of plant conditions as indicated by the ink charts on the control room walls. They too appeared okay, however, some of the ink charts were not recording.

Jim found this shift as busy as any other, checking off tasks, answering the phone, sending faxes, adjusting control panels and monitoring gas flow process on two separate monitors. As a part of the ordinary running of the plant, Jim called Ronnie on a few occasions, asking him to do certain things to enable him to run the plant.

So, the day shift of the 25 September 1998 commenced like every other shift. However, it wasn't too long before the routines and processes the workers had operated for years started to be interrupted.

After the toolbox meeting, Ronnie went to the control room to start his shift, which normally started with him going to his assigned area with the Portable Data Terminal to check and take readings of the equipment.

However, before he started, Jim asked Ronnie to close the TRCB3 bypass valve. When Ronnie got to the bypass he noticed an operator's lock and chain dangling on the valve but not locking the valve. The red tag read "Bypass locked and tagged per shift supervisor". But the valve was open, so Ronnie decided to close it as he'd been asked.

Ronnie then started his daily route set by the Portable Data Terminal requesting data to be entered in a specific order. The route was comprehensive. Ronnie walked to the flash drums called GP1105.

There, Ronnie took a couple of pressure and level readings on each one, then the condensate deethaniser and the crude deethaniser, where he took the two pressures. Following this, he continued on to the lean oil reclaimer to take a temperature flow and a pressure reading. This was followed by the A & B debutiniser, where he took two temperatures of the reboilers; then the heaters, which he inspected and recorded the temperatures for, and so on. All in all, there were about forty stops and pieces of equipment to measure and record.

During the readings, Jim radioed Ronnie. He asked if Ronnie could open the bypass valve, which Ronnie did.

As Ronnie walked about the area, he noticed a product leak at the very bottom of the end of one of the heat exchangers referred to as GP922. There was a drip tray under the leak. Ronnie repositioned the drip tray before going back to the control room to tell Jim and the supervisor about the leak.

A leaking vessel was not unusual. It was something that the operators would come across on a day-to-day basis. However, what was unusual about this leak was that it was in a vessel that the operators hadn't seen leak before.

Puzzled by this, the supervisor and Ronnie left the control room to observe the leak.

Shortly after, Jim noticed that one of the lean oil booster pumps, GP1201, had shut down.

He radioed Ronnie. "Ron, the pumps on GP1201 have shut down. Can you restart them?"

Ronnie went to the pumps and tried to restart them without success. He was baffled as to why they wouldn't restart.

He returned to the control room and asked Jim, "I couldn't get them to restart Jim. Do you think they've shut down because of the low level in GP1110?"— this was the oil saturator tank. The particular level Ronnie was referring to was the lean oil liquid level. A check revealed that there was a low level in this tank. Consequently, another pump, the GP1202 pump, shut down as well.

Shaking his head, Ronnie left the control room to try and restart both pumps. He hit the restart buttons which are located about one metre off the ground adjacent to the pumps but neither one of the pumps started. It was obvious to Ronnie that the level of the lean oil liquid in the pre- saturation tank was low and that the pumps had their low level shut downs activated. Ronnie rubbed his chin. He wondered why they were having such low levels.

In the meanwhile, Jim attended to the rest of his duties, such as adjusting the inlet gas flow to Gas Plant 1 to help cope with the process situation at hand and to prepare for the upcoming gas day.

One of the alarms started to sound and light up. It indicated the shutdown of the rich oil demethaniser reflux pumps referred to as GP1203A and B. The pumps would have shut down due to a low level in the reflux drum. But when Jim looked at the level of the reflux drum shown on the ink chart, he saw there was sufficient level for the pumps to continue running. He couldn't understand why they had shut down. Jim called Ronnie again, "Ronnie, GP1203 A and B have shutdown. Can you restart them?"

When Ronnie went to the pumps he found that one pump was still running and achieving a small flow. He went back to the control room and told Jim "I'll go back and keep trying to restart GP1201 and GP1202."

Ronnie looked at the pipework and the area around the rich oil demethaniser. He noticed that heat exchanger, GP922 had icing on the west end and GP905 had icing on the east end. Then, he saw icing on the rich oil demethaniser, the suction and discharge lines on the very north pump of the three GP1201 pumps.

He said softly, "In the eighteen years I've been here, I've never seen icing like this. I wonder how long it was between 1201 pumps shutting down and the ice appearing."

It was baffling. Ronnie walked up to the vessel, stood in front of the far end of the exchanger and rubbed his hand across it. He could see and feel the ice build-up on it. Ronnie muttered, "What the hell is going on here?"

These problems were occurring while Jim was attending to the daily tasks assigned to him as a control room operator. The reality was that despite the demands of the equipment failures, they comprised a small fraction of Jim's other responsibilities. These included the radio and phone conversations, attending to alarms, monitoring equipment, administration tasks and so on.

Jim was also involved in critical function testing being conducted at the Westbury Pumping Station near Trafalgar. The testing involved a maintenance technician at Trafalgar triggering an alarm which would activate in Gas Plant 1 control room where Jim was. Jim would have to answer that alarm and contact the technician at Westbury to tell him which alarm went off and whether it had reset.

While attending to the functions in the control room, Jim saw that one of the pumps had restarted and shut down again. He radioed Ronnie again and said "GP1202 restarted for a while but it sucked the level straight out of the OST (referring to the Oil Saturator Tank). We'll have to wait for the level to build up again before we restart it."

Ronnie replied, "I only got GP1202 started, but I couldn't get the GP1201 pumps started at all."

Ronnie asked, "Jim, what level is the rich oil demethaniser at?"

Jim wasn't familiar with that indicator or whether there was such an indicator in the Gas Plant 1 control room. He looked along the control room panel for the indicator, but he couldn't locate it and was too busy to spend much more time looking for it.

Jim replied, "I can't find any indication of a level in the control room, Ron. And I'm run off my feet. I'm flat out. I don't have time to look for it."

"Oh, OK. I'll go and check out the level on the ROD itself." When Ronnie got to the rich oil demethaniser, he couldn't see the level — not even with a sight glass.

In the control room, Jim felt the intensity of the situation. The events that were taking place were not a regular occurrence. The process upsets, alarms and phone calls meant that Jim needed additional help with the process issues and the restarting of equipment, so he called the supervisor for additional workers to help.

"Jim, I'm in the production meeting and the day crew are all on training, but they'll be back shortly," the supervisor replied.

By now, a gas-fired reboiler, GP502B, associated with the crude deethaniser tower, had also shut down. Jim was very concerned.

Ronnie called Jim and asked "Jim, what heater did we lose? I'll go and have a look at it."

Jim replied, "502B. Thanks, Ron."

However, when Ronnie got to the vessel he could not relight it because it needed a sufficient level in the tower for the pumps to operate before the relighting could take place.

Ronnie radioed Jim and asked, "Have we got enough level in that tower to run the pumps? I started the pumps, but they didn't operate."

Ronnie was asking whether there was enough level in the crude condensate deethaniser tower to run the pumps which fed the GP502 reboilers.

Jim replied, "Yeah, we did have plenty, hang on, I'll sort out what's going on here."

The level in the crude condensate deethaniser provides the product feed to the heaters via the pumps. When the heaters shut down because of low level in the crude condensate deethaniser, this shuts down the pumps which feed the heaters. To get the heaters back on, a sufficient level in the tower has to be established to start the pumps and feed product to the heaters. The red illuminating alarms for the crude condensate deethaniser low level shutdown, the 502B heater low-flow shutdown and the low- pressure fuel-to-heater shutdown were sounding simultaneously.

Jim replied, "The level in the crude deethaniser tower has just dropped, if you relight GP502B it'll shut down again. It'll have to wait until the level gets back before we get you to follow that up."

The alarm for the CSP control panel sounded and lit up. This time one of the technicians responded to it. He asked Ronnie to open the bypasses on PRC10 (the pressure recorder and controller), which controls the pressure at the top of the product debutaniser tower in the crude oil stabilisation plant.

Ronnie radioed Jim to ask if there was sufficient level in the crude deethaniser tower to start the pumps. "The pumps are not running, Jim, they're still in test mode. I'll go and see if I can start them up."

Ronnie made his way to the pumps. He managed to get GP1201 running. Then he tried to restart GP1202 but it didn't respond. He tried a second time and managed to start the second pump.

Ronnie sighed as he went to LC8B, which controlled the rich oil level in one of the absorbers, to check the levels. He wondered if the icing on the heat exchangers GP922 and GP905 may have been a sign of the low levels. It was hard to gauge the levels visually so Ronnie used a sight glass. Even then the level was hard to see.

Ronnie contacted Jim. "Jim, can I open the bypass valve to hear or feel any liquid or gas passing through? I can't see the level, mate."

"Yes," Jim replied.

Ronnie opened the bypass valve and determined there was liquid flowing through.

Almost two hours into this busy shift, Jim contacted Ronnie to let him know that the Longford Liquid Recovery, the LLR, plant shutdown was commencing. This shutdown is indicated by several Bailey alarms that light up and which are silenced manually. The LLR was near the GP502B heater and Jim thought if Ronnie was close, he could help out by restarting the LLR.

"If you're still at GP502B, you'll be close to the LLR plant. Can you restart it?" Jim asked Ronnie.

Ronnie replied. "I'm not at 502B, Jim. I'm at the LC8B."

"Oh, I see," replied Jim. "I'll get someone from day crew to do it as soon as they're available."

Numerous controls were in place throughout Gas Plant 1 to regulate the levels, temperatures and pressures in various vessels. Among these were the level controls for the condensate in the bottom of Absorber B and for the rich oil in Absorber B. They were known respectively as LC9B and LC8B.

Sometime later, the plant supervisor asked Ronnie to open the level control bypass at Absorber B. Ronnie radioed Jim. "Jim, I've opened the LC9B bypass."

"Righteo," replied Jim.

Ten minutes later, the plant supervisor called Ronnie from the control room. "The level's starting to come back in GP1105A," referring to the condensate flash tank. "You might even crack the bypass a bit more around LC9 on B."

"Yeah, I'll do that," Ronnie replied.

"Okay," said the plant supervisor.

While Ronnie was standing around the 1201 pumps, he smelled something. He thought it smelled like a 'hot smell' coming from somewhere. Ronnie looked around and up and noticed that GP1202, which he had started, was hot and billowing smoke. He thought the pump was going to catch fire. He tried to shut it down by pressing the stop button, but it would not shut down. Ronnie thought to himself, *these discharge pipes run at zero degrees. They're cold pipes, why on earth are they cooking?*

Immediately, Ronnie radioed the plant supervisor to take a look at the pump. When the supervisor saw the pump, he tried to stop it by pressing the stop button too, but it didn't work. Then he went to the switch building to trip the breaker.

Ronnie was worried about the possibility of the pump catching fire. He called Rob Miller and asked him to help him unwind the fire hose and pressure it up just in case the pump caught fire. Then, he radioed the supervisor and asked, "Did you trip the breaker?"

The supervisor replied, "Yes, I did, Ronnie."

Over the next few hours, the situation escalated and more and more workers from different parts of the workforce got involved in trying to rectify it. Jim, Ronnie, the supervisors, other operators, electricians and maintenance personnel focused on restarting the lean oil pumps.

The plant supervisor went back to the control room and spoke to one of the electrical supervisors about the gas pump, GP1202. Shortly afterwards Ronnie informed the supervisor that the leak from the heat exchanger GP922 was getting worse.

The supervisor motioned with his head. "Let's go and check the leak, Ronnie," he said as he walked out of the control room with Ronnie close behind.

To their dismay, both ends of the vessel were leaking.

The plant supervisor rubbed his chin and said, "We'll need to shut down Gas Plant 1. I'll call the control panel operator in Gas

Plant 2 and ask him to prepare the plant to receive the KVR gas," referring to what is properly described as crude pipe vapours. Then he contacted Gas Plant 1.

One of the technicians answered on behalf of Jim. "Mate, shut down the rich oil flash fired reboilers, GP501A and B." The GP1204 pumps were manually shut down from the control room.

The plant supervisor then went out and isolated GP922 from the rich oil demethaniser by closing one of the two Level Control 10 block valves.

Ronnie sighed. He knew shutting the gas plant and redirecting KVR gas meant not only stopping the flow of inlet gas from offshore but also transferring the gas coming from the CSP via the KVR compressors to Gas Plant 2. The transfer involved opening and closing valves in Gas Plant 1 and Gas Plant 2 to redirect the gas.

The valve handle was about thirty-six inches round and took two people to open fully. Rob Miller was the area operator for the KVR. Ronnie and Rob worked on the redirection of the gas to the other gas plant, which took about forty minutes.

Ronnie said to Rob, "It'd be good if we could get some help, but there's no one. They've all got their own work to do."

Rob nodded silently as he helped turn the valve handle.

Jim called Ronnie. "Ronnie, as a safety precaution, can you block in the fuel gas block valves for the fired reboilers?"

Ronnie replied, "Yeah, mate. I've got to do the transfer of the KVR gas to Gas Plant 2 first."

About half an hour later the transfer was complete.

Shortly thereafter, the plant supervisor noticed that Gas Plant 1's crude oil stabilisation plant electrical tie had been lost due to the shutdown of Gas Plant 1. Rob and another area operator went to the old generator building to restore the tie.

In between time, the plant supervisor went to the offices to get help. He and the relief plant supervisor started walking to Gas Plant 1, stopping at Peter Wilson's office first. By default, Peter, who was the maintenance superintendent, was the most senior person at the plant on this day. The position of operations superintendent was vacant and was being filled by an acting superintendent. He was on leave and the person who was to relieve for him was away due to illness.

Peter listened intently and once the two plant supervisors had left his office to continue on their way to GP922, Peter rang the plant manager to let him know about the leak. The plant manager asked Peter to call the operations manager in Melbourne to have the matter investigated. He too was not available, so Peter left a message for him to call back.

When the plant supervisors arrived at GP922, they observed a pool of liquid under and around the vessel and at both ends, although at this time the leakage was less than it had been. The area of the spill was estimated to be approximately six metres in length by over three metres in width, with a depth of ten centimetres, about half the liquid was in the stones underneath the vessel. They estimated that there were over one thousand litres of liquid on the ground.

At one point, Jim asked Ronnie if the bypass on the level control valve on Absorber B was open. Ronnie said he'd have a look. He asked Jim, "Do you want it open or shut?"

Jim replied, "Close it," and then "we're back in control here."

By this time, other operators were having trouble with the propane pressure in the Gas Plant 1 refrigeration system. The pressure was low and for that reason Jim asked one of the operators to shut down about half a dozen fans in the KVR area.

Jim also contacted Ronnie and told him that there was a low-level shutdown on the condensate deethaniser reboiler GP502A, so it needed blocking in. Ronnie replied that he was at Absorber B and that the valve was wide open and that he was blocking it in. Jim told Ronnie to let the level controller do its job and that it should be right so long as the bypass was shut.

"I don't think we've got a level in there, Jim," Ronnie said.

Jim replied "Previously we've had 105%. Now we're controlling it at 56%."

"OK then, I'll leave it," said Ronnie.

Soon the temperatures in Absorber B started falling. Jim asked Ronnie whether the temperature control system bypass was open.

"Yeah. It's open," Ronnie replied.

While things were still uncertain, they seemed to be under control. The workers were weary and they had been running around continuously since the start of the shift.

Ronnie turned to Rob and sighed as he said. "I'm going for lunch mate. Everything's been shut down and looks like it's under control."

Rob nodded and replied, "Yeah, I'll go for lunch too."

As Ronnie ate his lunch he thought, *from the time Jim and I started the day shift until now, this morning has run in cycles from one problem to another.*

Mentally, he replayed the morning's events. Alarms sounding, temperatures falling, the operators contacting Ronnie and indicating where the trouble was, asking him to open bypasses, Ronnie in turn asking about the levels of each equipment and trying to rectify each situation according to its remedy. And the ice on the heat exchangers, the ice, that was the most baffling thing.

As Ronnie thought about the day's events, two electricians walked up to him. One of them said, "Ronnie, GP1200 nearly caught fire! Can you rap the breaker out in the switchgear building?" The breaker was like a large fuse.

Ronnie rushed to the switchgear building, pulled out the breaker and attached the right locks and tags to stop it being used until it was tested.

Then, he joined a group of people that included Peter Wilson, John Lowery and a few operators. They were discussing the irregularities of the day. One of those was the leak from GP922.

The liquid that had leaked from GP922 was being collected in drip trays, but these had overflowed. John Lowery, who was the acting maintenance supervisor, was asked to arrange for maintenance staff to re-tension the bolts on the leaking flanges. At the same time, three operations personnel from the day crew pumped the spilled liquid into the open drain system which is a pipework system used for the safe disposal of liquids. John arranged for one of the mechanical technicians to investigate the spill and for a couple of maintenance fitters to attend to the re-tensioning of the bolts.

When the mechanical technicians arrived at GP922, they estimated the spill to be over one thousand litres. Peter, however, estimated it to be over three thousand litres.

The maintenance fitters checked the tensioning on both ends of GP922 and found that it was within specification. John also asked them to check the flange on one of the pipes on the top of GP922. It was not leaking and it was tight.

However, one of the maintenance fitters noticed that the east end of GP922 was completely frosted up, as were some of the connecting pipes. The frost was one centimetre thick.

By this time Peter, the production co-ordinator and a few others, including the construction site superintendent, were in the control room with Jim Ward. They had a general conversation about what was happening at that time and they observed the Gas Plant 1 processing conditions on the panel.

"There's no gas flow or lean oil circulation. The temperature in Absorber B is about -35°C; the pre-sat vapour tank is empty; the rich oil flash tank has a low level at 26%; GP922 heat exchanger is leaking; and the reboilers are shut down. The ROD and its reboilers have to be about -48°C," one of them recounted.

Suggestions were made by various people that there could be a control problem or a remote possibility of the formation of a hydrate. But no one knew. Soon after the discussion, they left Jim to get on with his job in the control room and each went to address the issue relevant to them.

Back in the office, Peter told the production coordinator that there was a leak on the heat exchanger. "Come with me so we can assess the volume that's been spilled. We might have to report it, depending on the volume of spillage," said Peter.

Regulations required that any hydrocarbon leak greater than two hundred kilograms was to be reported to the Country Fire Authority.

Peter and the production coordinator went to the maintenance fitters from Gas Plant 1 control room. On his way, the production coordinator noticed that the feed line to the rich oil demethaniser was frozen. It was covered with white ice. The ice covered the whole length of the pipe, which ran from the pipe rack at the top of the control room area to the tower. The production coordinator arrived at the control room where he was told that the rich oil fractionator reboilers were off line and that attempts were being made to establish lean oil circulation.

From the control room, the production coordinator went to GP922. After a conversation there with Peter and John, he concluded that the leak from GP922 was not significant at that time.

He watched the plant supervisor and acting plant supervisor open the Level Control 10 block valve to allow rich oil to return to GP922 and then back to the rich oil fractionator.

He returned to the control room where Jim pointed out that the temperature control system to GP903B recorded a temperature of -48°C and that the temperature was still dropping.

Two operators left the group to go to help start one of the pumps, GP1204. Before doing so, one of them radioed Jim to tell him what they were going to do. Jim told him to go ahead because it was necessary and that several pumps would not run unless this particular pump was running.

After checking that the valves on the suction side of the pumps were open, the operator pressed the start button of the pump. The pressure rose as if the pump was pumping, but there was no flow.

The production coordinator radioed the relief plant supervisor from the control room. He instructed him to put two particular flow control valves into bypass. This was the flow controller that regulated the flow of lean oil through the rich oil fractionator reboilers. In that way, a flow was established through the reboilers.

Then, they discussed whether a specific pump, GP1204, should be started. The production coordinator said he wanted a flow in both the reboilers, which had not been re-lit, because they had a large radiant area in them which would drop the temperature of the circulating lean oil.

Jim told the relief plant supervisor that there was a flow in both heaters and that was the way they wanted it. The control room charts showed that the flow had restarted.

Meanwhile, Ronnie was speaking with the production coordinator. He said, "I don't know what's happening. We don't even have time to step back and try and work it out. Things are happening too quickly, and things are happening that I've never seen before. What's going on?"

The production coordinator shook his head. "No bloody idea. It's been like this for hours and we don't know what's happening. Can you go down to the plant and check the pump?"

On the way out, Ronnie noticed John Lowery standing nearby. Impulsively, he turned to him and patted him on the belly and said, "I'll be back in a minute mate." Little did he know, that would be the last time he would see John. Ronnie got on a pushbike and rode in the direction of the area he had been asked to go.

The maintenance fitters who had been re-tensioning the bolts on the GP922 flanges finished their work without making any significant changes to the bolt tensions or the rate of leakage. There was some discussion about whether the leaks could be stopped without replacing the flange gaskets and whether GP922 should be isolated and depressurised to allow this to proceed. In the end, it was decided that the best method of stopping the leakage was to warm the vessel slowly by restarting the flow of warm lean oil. The plan was to bring the lean oil back into circulation, but to leave the some of the heaters off so that the oil would be relatively cool.

By this time, the heaters had been off for an hour and were not re-lit. The thinking was to circulate the lean oil through them in the hope that their large surface area would provide cooling for the lean oil and thereby reduce any thermal shock.

The maintenance fitters finished their work and one pump started up and then another. Several of the operators were stopping and starting the GP1201 pumps. Eventually, both GP1204 and GP1201 pumps had started. One of the supervisors then went to the LC10 block valve and opened it to allow flow to GP922. Immediately, the vessel started to leak again; this time at both ends. Liquid was fanning out one metre, in a semi-circle from the bottom half of the heat exchanger head. Some of the men left the area because they sensed the danger in the situation while others remained in the vicinity.

In the meantime, Jim and the production coordinator were looking at the charts in the control room and observed that there was no change in the level in the oil saturator tank. Jim radioed the operator. "There's no flow through the GP1201 pump that's been started," he said. "Just swap the booster pumps over for us."

Two operators returned to the area and saw that GP1201, which had been started only a few minutes before, had stopped. One of the operators walked to another pump and pressed the start button to try and restart it. Even though there was no flow, the pump continued to run giving the false impression that there was a flow.

Peter was down at the plant when he was asked to call back the operations manager whom he had left a message for earlier in the day.

He went to the control room to make the call.

Jim was standing in front of the panel and was too busy to notice that Peter had come in until Peter spoke. Peter and the operations manager discussed the situation at the plant. Jim continued working as he overheard Peter giving a run down on what they were doing about the situation. He said that he had been asked by the plant manager to notify the operations manager of the lean oil leak onto the stones from the head of the GP922 heat exchanger. He also said that they were tightening the head and cleaning the spill with pumps and absorbent materials. By all accounts, the situation seemed to be under control and didn't seem to pose any danger at the time.

"Do you know what caused the leak?" asked the operations manager.

Peter explained that the shift felt there was a problem in the absorbers, maybe a block, possibly a hydrate, causing a loss of the rich oil in the system. Peter explained further that they were trying to find out more.

"Also," he added, "I called to let you know that Gas Plant 1 has been shut down. All the vapours have been redirected to Gas Plant 2. This won't affect gas sales because the order can be met by Gas Plant 2 and Gas Plant 3." Peter seemed keen to get back to the job, so the conversation ended with an undertaking to keep the operations manager informed.

Jim looked up at the clock on the wall; it was 12.10 pm.

In the meantime, the production coordinator left the control room to check GP905. When he got there, he saw ice on the areas where there was no insulation. Shaking his head, he realised that if the temperature was too low there was a danger that an impact could cause a brittle fracture. By this time, it was possible that lean oil flow had started.

So, he tried to minimise the flow of lean oil through GP905. He looked at the temperature recorder and controller 4, or TRC4, valve 1 and observed that it was closed. To obtain minimum lean oil flow through the GP905, this valve had to be fully open.

Back in the control room, after Peter got off the phone, he went to the south end of the control room. Just then, Jim got a radio call and as he answered it, he looked over to Peter who said, "I'd better get back to 922 and check the smoke there." Jim nodded, acknowledging Peter's comment and departure. The door was left open after Peter left.

Jim answered the radio call. It was the production coordinator who asked him to close TRC4 so as to open valve 1. Jim misheard the instruction. He thought he'd been asked to close PRC4 not TRC4. PRC4 was the pressure controller for the rich oil flash tank. Because of this misunderstanding, Jim made no adjustment to the TRC4 controller position and the controller output remained fully open.

The production coordinator waited to see TRC4 valve 1 open, which he expected would happen when Jim carried out his instruction.

But nothing happened. He was confused, even when Jim confirmed over the radio that the TRC4 was closed. Not realising he and Jim were mishearing one another, his confusion was compounded.

He looked at the field switch which converts the valve configuration for the GP905 heat from demethanising to deethanising mode and he switched it to the opposite setting. The reason for doing this was to try and get the TRC4 valve 1 to open to minimise lean oil circulation through GP905, although he was not sure what the outcome of operating the switch would be. It did not appear to affect valve 1.

The production coordinator radioed Jim again. "Jim, go for maximum output on TRC4," he instructed as he looked at the GP905 exchanger and saw no change in the level of icing. Then he looked around and saw valve 1 opening and the steam rising. He thought it may have just been a delay that valve 1 appeared not to respond. He looked at the three-way valve. It did not appear to have changed. Then he stepped back to about the centre of the GP905 and looked from end to end to see if there had been any change.

The insidious situation deceived the workers into believing that they could conquer the day. One by one they continued to attend to their duties, routines and work at the plant.

One stopped to have a smoke at the smoko shed to reflect on the madness of the events so far. One mounted his bicycle to go to lunch while others were coming back from theirs. Some persevered with different malfunctioning vessels while others looked for solutions elsewhere.

The one thing they all had in common was that no one would be spared from the onslaught of the calamity to come.

Rob mounted his bicycle, rode to the canteen and bought a fish burger for lunch, before going to the lunchroom in the control room to eat it. As he ate the burger, one of the maintenance crew came to him and told him that there was another piece of equipment that had shut down and they could not restart it. Rob radioed his supervisor who organised someone to attend to this issue.

Soon after, while finishing the last piece of the fish burger, Rob walked out of the lunchroom to go to the troubled area the maintenance crew had told him about.

Just as he stepped out, he heard a huge rush of gas which sounded like a jet engine or an aeroplane. He spun around wondering what happened. His mind was ticking fast. He saw smoke and a thick cloud rise from the area the noise came from. It resembled a very foggy day and it was travelling away from him. It was roaring. Then the fire alarm sounded. Rob stood there thinking, *this shouldn't happen here.* He was in shock and he was confused. He just stood there thinking the same thing over and over, *this shouldn't happen here.*

Then he saw a man run out of the same door he had come out from, he was running to a monitor, which is a fixed position firefighting nozzle used to spray water or foam onto a fire. Rob started running to the monitor too. They reached it at the same time. One of them positioned it on the cloud that was still billowing and the other turned on the water supply. Then they both ran towards the direction of the cloud to see if anyone was hurt.

Rob saw someone covered in a black substance from head to toe, standing hunched over and looking from side to side. He did not recognise the man, until he got closer and, seeing the man's face, he realised it was one of his workmates.

Rob told the worker to move away from the area because it was not safe.

Next, he looked towards the gas cloud and saw another figure covered in black struggling to stand or walk. He was tripping and stumbling. He had a cut over his eye and it seemed that he was bleeding from one ear. Rob grabbed hold of him to try and help him walk to the control room. The man refused Rob's help. No matter what he suggested the man did not cooperate. Rob wondered if the injured worker could not hear him because the blast had affected his hearing. For a few seconds the man did not say anything. As he stood there, hunched over, he noticed a pool of blood on the floor. He looked up at Rob and asked, "Am I bleeding?" Rob nodded and said "yes".

Once he realised he was bleeding, the man reached out to Rob as if agreeing to get the help Rob had been offering. Rob helped the man walk to the control room and sat him in the panel operator's chair. He turned to someone in the room and said, "Can you keep an eye on him? I'm going to see if there's anyone else out there."

As he walked to the door, unexpectedly, there was another abrupt and loud explosion. Rob saw a huge gas cloud ignite and catch fire. It made the sound of gas roaring. The control room shook even though it was built on concrete blocks. Rob thought the walls were going to collapse. Instinctively, he squatted down and looked towards the wall, which did not collapse.

He stood up and rushed outside to look for other injured people and to make sure the fire extinguisher pumps were turned on. There were several other explosions. He was running out of breath because he had been running and breathing in the smoke and other substances released by the explosions. He stopped for a few seconds to try to catch his breath, then he noticed someone standing nearby and told him there were other fires. Together, they ran to those fires to try to extinguish them. As they ran, Rob noticed a safety helmet on the ground. It was mangled, Rob wondered who it belonged to.

They ran to some fire extinguishers which were covered by plastic intended to protect them from the elements. They grabbed an extinguisher each. But the flames and the heat had melted the plastic covering and shrunk it around the extinguishers and into the nozzles where the foam was supposed to come out. They wouldn't bloody work! So, they grabbed other extinguishers that did work.

They ran around squirting small flames to stop them spreading. Then there was another explosion which inflamed the small fires. They agreed they could not do much there, so they parted ways. The other worker ran towards Gas Plant 2 and Rob ran towards the major fire, to the control room. He helped injured people get out of the control room into the first aid room. When he got to the first aid room, everyone was scared about the close proximity to the fire. So, he helped with the evacuation of the first aid room too.

All the while, Rob's heart was pounding. He wanted to get out of the plant. But he felt responsible for his workmates, so he stayed. He was in shock and wondering what the bloody hell was going on, what had happened. It was still perplexing to him. The last thing he had seen before the explosion was the leak from the head of an exchanger and then the explosions. It was confusing. Rob could not work out what had happened.

Having helped evacuate the injured, Rob ran from one area to another working the fire pumps and making sure the pressure on the extinguishers was right.

As he did this, he heard a head count over the radio. It was obvious that a couple of people were not accounted for, including Peter. Eventually, an order was given to evacuate.

At about ten minutes past twelve, Marty headed off for lunch. On the way up he passed John Lowery, who was walking along the footpath with another technician. They were discussing something so Marty passed them by and kept walking to the old guard house up near the cannery. Having finished his lunch, Marty was resting before he went back to work. Suddenly, he heard a loud 'boom'! Initially, he thought it sounded like a compressor backfiring. A sound he was familiar with.

Shortly after, there was a bigger sound and the entire place shook. It was as if someone had dropped a bomb in the plant. The plates, knives, forks and windows rattled in the canteen and everything suddenly went quiet. No one in the canteen said a word. Marty and a few workers got up and rushed to the south side of the canteen. They looked down to the control room, but they could not see it because other buildings obstructed it. Someone said, "I think it's the control room." So, Marty and a few others ran to their bikes and rode them in the direction of the control room.

When they got to the control room, they saw the fire that looked like it was amongst the exchangers and pipe works.

Marty assumed there would be people injured. One of the workers was setting up a firefighting monitor, due east to the control room, about fifty metres from the fire. Marty and another worker raced over to him and took control of the monitor to help him. The fire was high and intense. The entire area was hot. The water wasn't making the target. It was steaming off the floor before it could hit the fire.

The worker who was on the monitor went to attend to the fire in the fans at the back of the control room. He grabbed a dry chemical fireball which was a bottle on wheels and weighed about fifty kilos. He was dusting the fire down with dry chemical, trying to put the fire out.

Marty suddenly realised that they were standing on stones at the monitor without any protective clothing. Marty shouted, "We've got to get out of here!"

They left that monitor and ran between the fire and the control room to another monitor which was about thirty metres from the fire. They re-aligned that monitor as they had been trained to do, to keep integrity of the other vessels intact. The rationale was to concentrate on keeping the pipe works and vessels cool instead of trying to put out the fire, which was impossible.

After positioning the monitor, they ran along the footpath directly behind the control room to get another fireball. Marty wheeled the bottle and the other worker carried the nozzle. As they were running back towards the fire to help the other worker dust the fire down with dry chemical, Marty saw the workers suddenly run off.

He felt his workmate grab his arm and yell, "Let's get the fuck out of here!" Marty realised that there was another explosion. They sprinted away from the fire towards the back of the control room.

After this explosion, Marty and his workmate got more equipment and ran back to the fire. They ran out hoses, then they ran back towards the fire to put a monitor on the vessels in order to keep them cool and intact. Then they hooked up the hydrant. All the while, Marty felt isolated and lonely, as if he and his workmate were the only two people in the world. Marty looked up and saw another worker running towards them. Suddenly, he felt less isolated, as if the worker had brought hope.

They hooked the hose up to connect the crossfire monitor and ran it back down the footpath. Marty had been trying to hook it up but he couldn't. He kept thinking, *it's a right-hand thread so why can't I grip on the thread?* Then he realised it was because his hands were shaking. Eventually the thread hooked up, and they opened the valve.

Suddenly, they saw a supervisor running from behind the switchgear building. He was yelling and screaming. Marty raced over to him to explain what they were trying to do. The supervisor kept yelling, "We have a man down! We have a man down!"

Marty's stomach dropped from the shock of hearing these words. The supervisor grabbed Marty by the shoulder and beckoned him to follow. Marty ran as fast as he could to the back of the switchgear.

When they got there, Marty saw the man the supervisor had told him about. He looked like a twisted mannequin lying in the stones. His overalls were black. They were like tissue paper even though they had fire retardant in them. His overalls on his right leg were opened and he had a ball of flesh sticking out of his thigh. It resembled a cauliflower. Marty wondered why he couldn't see blood. He thought *well where's the blood? Why isn't he bleeding? He's got a horrendous chunk of meat sticking out of his leg, why isn't it bleeding?* Then it occurred to Marty that the man was scorched. That part of his leg had been cooked. That's why there was no blood. His flesh had been seared. The realisation was sickening.

Marty felt distressed by the heat, the constant noise and what he was seeing. Instinctively, he grabbed the man on the ground by his ankles as if to drag him out. The worker looked up at Marty with terror in his blue eyes set in his blackened burnt face; an image that would be imprinted in Marty's mind. He said, "Watch my leg, Marty."

Marty stopped and thought, *who the hell is he? Maybe he's one of the contractors.* He knew that there were some contractors from a hire company and they wore navy blue overalls. It appeared to Marty that the injured worker was wearing overalls like the contractors wore. As the fallen worker was speaking, he appeared to have no teeth. Marty's first reaction was to roll him on his side and clear his airways in case he was choking on his own teeth. The stinking heat was unbearable, it was just bloody horrendous.

Marty was still holding onto the man's ankles and the supervisor kept saying that he had a broken leg and that they were going to need a stretcher. Marty refused to leave the injured worker to go and get the stretcher. They were surrounded by flames, explosions and roaring pipes. Marty looked at the supervisor and said, "Let's get him out of here."

They grabbed the man from underneath the knees and what remained of his overalls. He kept pleading, "Watch my leg, my leg's broken."

They tried to carry him, but he was too heavy. It took a lot of effort for them to drag him behind the switchgear, away from the radiant heat. Once he was out of the direct flames and scorching heat Marty ran to the control room to get a stretcher.

Ronnie was riding the pushbike across the plant yard, about twenty metres from the heaters, when he heard a huge thump. He felt it in his entire body. He stopped the bike and dismounted to look towards the direction of the sound.

When he looked around he saw a white cloud of vapour. The cloud of vapour stealthily advanced towards him, as if pursuing him.

He got on his bike and pedalled away from the cloud. As he rode the bike, he looked towards the heater area. He saw a clear, red flame, as high as twenty metres, mixed with a huge cloud of dust, stones and debris shoot up into the sky.

He changed the direction of the bike and rode another twenty metres when suddenly he heard a loud explosion — the flames were an angry red and black in colour.

Petrified, he asked himself, "My God, what happened there? Who was there I wonder?" Then he thought about his wife and children.

Next, the explosion came down towards the heaters. Ronnie saw the cloud rolling towards him. Even though two of the heaters were off, there were still three others, which were a part of another process, that were still alight. The cloud started rolling down. It went up thirty metres or more into the air and across the gas plant and came down south towards the heaters.

As he stood there wondering what was happening, another explosion took place closer to him. Ronnie pleaded, "Oh God help us!" Then the realisation hit him.

He yelled into the radio, "Some of the other heaters are still on! Jim! Jim! Shut your heaters in! Shut your heaters in! Shut your heaters in! Something's happening! I think the explosion's going to the heaters!"

He turned back towards the loud explosive noises and noticed the cloud of eruption heading towards him. "Oh, God help us!" he shouted as he jumped on the pushbike and sped away from the cloud.

Then he saw the cloud ignite the gas on the very north-west corner and roll back. It only took a few seconds to roll back from about fifty metres and then there was a huge kapum! It was an explosion of angry red fire and black smoke. Everything around him seemed to go up in flames and smoke. The thick blanket of blackness blotted out everything.

Furiously pedalling, he turned back and saw a fire ignite the gas and roll back and explode. Ronnie kept thinking of the men — he wondered who was where? How could he save anyone, including himself? He kept looking around for a clear path to get back to the plant. Whenever he took a path he thought was clear, it was blocked by fire, smoke and small explosions.

Feeling lonely, helpless and frightened, Ronnie felt the terror. His breathing was getting faster and faster. He kept repeating to himself out loud, "It's happened! It's happened! We never thought it would, but it's happened. We have an inferno. God help us! God help me to get to the boys. I don't know who's there, but I know the boys will be trapped." As he repeated these pleas, he persevered from path to path, seeking a way to get to his workmates.

Again, he thought of his wife Rhonda and the children.

Ronnie was in shock. He felt the fear deep within. Even though he had known they worked with a volatile product, he never thought anything like this would ever happen.

It had happened! They had an explosion. They had a fire. Ronnie thought, *somehow I gotta get up through to where the boys are.*

He looked around and every route he took was blocked by walls of fire and smoke. He saw one narrow path and it too was engulfed in fire. *Crap! I don't care. I'm going through it to find my friends. I gotta get to Jim* he thought.

He pedalled as fast as he could and rode through the fire quickly to get to the control room in Gas Plant 1 where Jim was. Despite his own fear and exhaustion, Ronnie was determined to find his fellow workers and help them. Eventually, he made his way out and around the west of the plant and back into the north side.

When Ronnie got to the control room, he breathed in an acidic-type gas which blocked his breathing. One of the other operators told Ronnie to put the breathing apparatus on from the control room. Ronnie couldn't even get to the breathing apparatus. He could not breathe in or out. He ran back outside and forced himself to cough to get breathing again.

There, Ronnie noticed that one of the monitors had gone off the fire. He ran to it and swung it around towards the fire. But the water pressure was low and it could not reach the seat of the fire because so many other monitors were using water.

There was another explosion. It felt like it came at Ronnie. He ran out of its way and then he went back to the monitor. That's when he saw an operator walking up the footpath. He was holding his hand covered in blood. Ronnie thought it had been blown off. He ran across to the injured worker and held onto him to support him as they walked to the control room to get help.

Ronnie heard someone from the corner of the control room yelling, "Come and help me ... come and help me with this stretcher!"

When Ronnie looked back he saw it was one of the new trainee operators, Marty Jackson. Ronnie said, "I'll be there in a minute. I just gotta make sure this guy's OK."

Marty yelled back, "No! Come and give me a hand!" When Ronnie looked back, three or four other operators had gone to Marty's help.

Seconds after Ronnie got the injured man to first aid, he saw the operators bring in someone on the stretcher. Ronnie looked at him but couldn't recognise him. He was all black. Ronnie ran over to the operators and gave them a hand lifting the injured worker onto the table. Ronnie turned to one of the operators near him and asked, "Who the hell is this?"

"It's Adam," one of them replied. It was hard to believe. The injured worker looked nothing like Adam.

Then Ronnie rushed outside and saw one of the supervisors in a dreadful state. He had blood coming out of one ear and both eyes. He'd been hit with stones and covered by black soot.

Ronnie ran over to him and removed the man's overalls and boots before helping him walk to the canteen where the injured were being cared for.

When Ronnie went back out, he saw another man lying on the ground. He looked like he was having a heart attack. There was a paramedic with him. He asked Ronnie for help.

The paramedic said "He's hyperventilating. He can't get enough air. Help me get his overall off." Ronnie began unbuttoning the man's overalls and the paramedic tried to peel them off. But the man was fighting them and trying to put his overalls back on.

The paramedic shouted, "He's in shock! Keep an eye on him and make sure he doesn't get up!" The worker kept gasping for air.

The injured man panicked even more, violently protesting as he continued to gasp for air, all the while fighting Ronnie. "Take your hands off me! Take your bloody hands off me!" he yelled.

The paramedic yelled back at the injured worker, "Stop it! Calm down and breathe deeply."

The worker surrendered from sheer exhaustion. Ronnie peeled his overalls down to his waist. Suddenly, there was another huge explosion. The ground trembled and smoke billowed out from the explosion. The worker started to panic again. Gasping for oxygen, he pushed Ronnie away, as if trying to escape. Finally, overcome by exhaustion, he collapsed back.

The paramedic looked around and turned to Ronnie. "Give me a hand! We've got to get out and we've gotta get him out of here!"

Ronnie and the paramedic helped the injured worker stand up and walked him to the canteen to be treated until help arrived.

They settled the worker down on the ground. Ronnie looked around and all he saw were injured workers groaning in sheer agony. They were being attended to by paramedics and other less injured mates.

He looked out at the plant and saw the angry fires continue to burn. He wondered if they would ever get out of there; if he would ever see his wife and children again.

As Ronnie looked about him and imagined himself being consumed by the fires, once again he thought of Rhonda and their children. He'd been thinking of his family the whole time since the initial explosion.

His thoughts were interrupted when he heard the injured man say, "I wanna be sick," trying to restrain himself from vomiting. Ronnie turned and looked at him and gently said, "Mate, just be sick. It doesn't matter. Just be sick." The injured worker vomited over himself and on the ground he was laying on.

Eventually, Ronnie, the paramedic and three other workers placed the injured man on a stretcher and carried him down to the main gate and out into a waiting ambulance.

As they walked to the warehouse fifty metres from the main gate, the fires continued to burn. The rescue crews had arrived and ambulances were taking injured workers away.

The supervisor asked Ronnie to go back to the plant. As Ronnie took a step to go back, there was another huge explosion. It went up thirty or forty metres. Ronnie thought, *Geez, there's no way I'm going down there*. Then he saw several operators running out.

Ronnie looked around at the devastation and realised that all the firefighting training they had received was in controlled environments. Even though they were real fires, they were controlled and there weren't the explosions during the training.

Ronnie shook his head and thought to himself, *when the real thing comes and you're in an area where the heat is melting even the pipes and they're exploding all around you, it's different. It's just different.*

CHAPTER TWELVE

BILL SHORTEN HEARS THE NEWS

> "BAD NEWS IS BAD FOR EVERYONE."
> Jill Lepore

There was hardly a day when Bill was not visiting a workplace, meeting members and delegates, and management. He felt most satisfaction in his work when he was amongst the delegates and members.

Ten weeks after winning the ballot for secretary, Bill was visiting the Chef Oven factory in Brunswick where the AWU had several hundred members. During the meeting with the manager and delegates, Bill's phone rang continuously. His initial reaction was to let the call be diverted to message bank to avoid interrupting the meeting. But the persistent caller eventually won. Excusing himself from the table, Bill picked up his mobile phone and walked away to the other end of the room to answer the call.

The rest of the people continued with light conversation until he could join them again. Suddenly, Bill's face turned ashen. He combed his fingers through his light brown hair. The silence in the room signalled an ominous event he was being told about.

"Oh, my goodness," he said. "That's terrible. Are people dead?" The other people in the room looked towards Bill.

"It sounds catastrophic. Like a bloody war zone." He ended his phone call and walked back to the meeting table.

Packing his notepad and other items into his well-used brown leather briefcase, he looked at the people sitting around the table. "I'm sorry, I've got to leave. There's been an explosion at the Esso gas plant out at Longford. I have to get down there, I have to get close to the people and their families." Whilst he spoke to the people at the table, in reality he was verbalising his next steps. "I've got to get close to the people," he repeated as he walked out of the room.

CHAPTER THIRTEEN

GEORGE PARKER AND THE HOLIDAY

> "VICTORY BELONGS TO THE MOST PERSEVERING."
> *Napoleon Bonaparte*

While George enjoyed the shop steward work, he decided to remind Terry that the twelve months had come and passed, and George wanted to relinquish the role as had been agreed when he first accepted it. But first, he wanted to take a break and go away with his wife Judy and some friends on a fishing trip — a holiday they had been planning for months.

The short break had started with excitement and anticipation; a well- deserved break for all, especially George. At the end of the first day of the holiday, they stopped at a motel to rest overnight. While the others looked around and muttered polite comments about the place being "clean enough" and "good enough for a one night stop over", George declared that he didn't want to stay and rushing to the car, to the amazement of Judy and their friends, announced that they were heading back home.

As George drove along the long country road, no one spoke, especially not George. The tension mounted as everyone else attempted some small talk without George's contribution.

Judy, politely but clearly annoyed, looked at George and asked, "George, what's the matter, you haven't said a word for hours?"

In deep thought, George replied, "I don't know love. I just feel strange. I just can't explain it in any other way other than to say I feel strange. I've got a funny feeling in my gut." Then, as if trying to detract attention from himself, he continued, "Anyway, I didn't like that motel we just looked at. We should go back home and sleep in our own beds."

Judy, clearly irritated, replied, "I can't understand you George. First you change shifts with Ronnie, so we could have a longer time away, and now you're heading back without letting us even rest for the night. The motel we just looked at was fine."

George continued to gaze ahead at the road as if he could not even hear Judy. Then sombrely he replied, "Neither can I love … neither can I." An uncomfortable silence followed.

The silence was interrupted by the ringtone of George's mobile phone. Judy answered the call. It was their daughter Jane.

"Mum, it's Jane. Can I talk with Dad?" she asked. Judy handed the phone to George.

Her voice shaking, Jane spoke slowly and carefully chose her words. "Dad, I'm calling you to tell you there's been a huge problem at the plant. I don't really know what's happened, I just heard there's been a problem and maybe some accident of some sort. I heard there are six men missing. Sorry, Dad." George was silent again but this time with reason.

Judy asked, "Are you alright love?"

George answered, "I don't know. I'm feeling numb. I wonder what's happened. If six men are missing, which six? I have a knot in my stomach Judy. It's my shift. My boys could be hurt, love."

Judy looked at George, reached over to him and gently rubbed his shoulder. As if trying to comfort him, she said, "It could be someone else's crew, love. It could be the managers."

George replied, "Yeah, but the law of averages says that my crew is running the plant, so they're gonna be there. Surely nothing major could have happened. At the end of the day, if everything else fails, the shut-down and pressure safety valve will save the plant."

As soon as they arrived home, George rushed into the house, changed his clothes and drove to the plant. He called Terry and a few of the men from the plant. He heard different accounts of what had happened and the more information he heard, the tighter the knot in his stomach got. He was told that the men were evacuated to St Johns hall in Sale. Someone else said there were men missing and others had been transported to the hospital.

Finally, he arrived at the plant and found it empty. He felt a hollow in his stomach. He watched the flames, smoke and the destruction around him and wondered what had happened.

From a distance, he saw Terry walking towards him. Both men were in shock. Eventually, George broke the silence and said, "Well, let's go over and talk with management from Melbourne. We've got to talk about where we're going and what's going to happen after this."

During the meeting with management, one of them said that he expected people who work for Esso to turn up immediately to help rebuild the plant. George was taken aback. Uncharacteristically, he leaned over to the manager and said in a deep, calculated tone, "I really think you haven't thought much about what you just said or the trauma the blokes have been through. I'm telling you now, you'll get the ones that can turn up and those who want to work and they'll be enough."

Leaving the meeting, George turned to Terry and said, "Mate, I'm going to call Billy Shorten. The boys need the confidence from knowing the union's behind them. They need to know that no matter what they'll be asked to do from here on, we're with them and I represent them and their interests."

George knew Bill Shorten from a previous incident that had involved the unfair dismissal of one of the workers who had breached a minor safety issue. Management had been relentless in their stance that the worker needed to be dismissed. George had escalated the matter to Bill Shorten, who had impressed George with his intellect and ability to negotiate and connect with the workers.

George thought it was no coincidence that only ten weeks ago Bill was elected to State Secretary of the Union.

George called Bill on the mobile phone and learned that Bill knew about the explosion and was on his way to Sale. George explained the day's events, about the fishing trip and the strange feeling he'd had, and ended by saying, "Bill, I think these were the worst couple of hours of my life. I mean, we always knew that we worked in an industry where that sort of thing could happen. But we also thought that it wouldn't happen to us. I really believed that at the end of the day if everything else failed, the shut down and pressure safety valve would save us. It didn't. That's why I'm so shocked."

Bill listened solemnly, amazed, but not surprised at the depth of George's connection to the people on the plant. Then he replied, "George, you and I know that there are two ways to handle a disaster, the high road or the low road. The high road is to accept responsibility; the low road is to lawyer-up and start the blame game. We both know that Esso has a very tough industrial relations record. For the company, it's always win or lose. And we know they've won more than they've lost." George was silent. Bill continued, "And we both know they're adversarial. I'm not sure how they'll play this one, but it makes sense there's gonna be a legal battle."

George was conflicted. "Bill, I'm not sure about that. I can understand them being adversarial for wages and things like that, but this is an accident. There are men missing and others in hospital. When you see the plant, it looks like a bloody war zone. What's Esso gonna be adversarial about? Surely they'll look after the men and their families."

Bill sighed. The lawyer in him knew most organisations were not measured on integrity and honour. They were measured on profits and returns to shareholders. While he felt incensed by this reality, he knew that this issue would be resolved in the legal sphere. So, he had started to prepare for the legal battle he expected would be declared.

"George, I've been talking with the union lawyers and to Bernard Murphy. We need a plan to represent the members in whatever comes out of this and to support them over the coming weeks and maybe months." Then, Bill confessed his only comfort in this tragedy. "George, all I can say is that if it's going to be a tough road, I'm glad the union's got you to stand by the members."

CHAPTER FOURTEEN

LUC WILSON

"NOT ONE WORD, NOT ONE GESTURE OF YOURS
SHALL I, COULD I, EVER FORGET..."

Leo Tolstoy, *Anna Karenina*

Life was sweet for twenty-two-year-old Luc Wilson. He was young, handsome and athletic. He was dating a girl he liked a lot and had good friends he could count on. His brothers, James and Brett, had decided to move to Melbourne for study and work. Luc, however, had decided to stay in Sale and live with his parents for the time being. Twelve months ago, during his university studies in construction, his father Peter had told him about a company looking for someone with Luc's academic background to work for them. As it happened, Luc was looking for a construction company to give him the experience he needed to add to his university degree.

As a child, both his parents had worked for Esso. Peter was an operator at the plant and Locky was a secretary. Sometimes during school holidays, they would both be rostered to work, so they would bring the boys to the plant with them. The boys could ride their bikes near the main entry or run up and down the hallway where the offices were while Peter and Locky did their jobs. It was a family atmosphere and for the Wilsons it was a home away from home. Luc's parents were fiercely loyal to Esso. They loved the company and often defended it if anyone had a bad word to say. Luc understood why.

Years later, when Luc's employment with the construction company brought him to Esso, he was very comfortable in that environment. He knew the plant well. He considered himself lucky. His routine included seeing his mum and dad every morning, working hard throughout the day and seeing his father at lunchtime when they would eat lunch together and just catch up. At other times, it was just a quick hello and a joke or two.

This day was different. Luc had stayed at his girlfriend's the night before and by the time he got home in the morning, Peter had left for work. At work, Peter had been busier than usual, and they had not had the chance to catch up. Around midday Luc went to lunch with his supervisor. As they drove back into the plant, Luc turned to his supervisor and said, "I'll jump on the bike and ride to Gas Plant 1 control room to drop the permits off for tomorrow." The supervisor nodded in agreement.

As they drove around the corner near the maintenance workshop they saw an implosion. Luc looked at his supervisor and asked, "Is that normal?"

The supervisor shook his head and looked stunned. He said, "No, that's not normal. Let's get the hell out of here." Hurriedly, he parked the car and they ran towards the gate. As they ran, they heard more explosions and the heaters igniting. There were sirens blaring in the background. Luc looked back towards the control room. He saw a ball of fire gushing towards the location of the initial implosion. Luc realised how serious the situation was. As he ran towards the gate, he felt a thud in his heart, as if his heart had dropped. He sensed an ominous ending; a premonition about what was to come.

Eventually, they got to the emergency assembly area where other people had started to assemble. There was confusion and uncertainty about what had just happened. Everyone was trying to work out what was going on. The rescue team was trying to locate people by two-way radios. Luc heard them continuously call his father on the radio, "Peter Wilson, come back. Peter Wilson, come back." There was no response. He knew Peter was in and around the vicinity of the fires and explosions. He sensed the worst.

As the explosions and fire escalated, Luc and the others were told to keep moving further away from the plant because the fumes were getting worse. They all congregated around the car park. Then someone in the crowd said, "We probably need to move further away because if this thing blows, it will level so many kilometres." So, the crowd proceeded to move into another paddock beside one of the helipads. Luc felt numb.

Suddenly, it dawned on him that he had no way of getting back home. As he thought about this, his attention was diverted to a conversation nearby. Someone said, "They can't find Peter."

As Luc looked around to see who had spoken these ominous words, he noticed Marty Jackson nearby, clutching the gates with both hands like a prisoner. He saw a haunted look on Marty's face as if he was transfixed by the horrors of what he had just witnessed. Luc knew he would never forget that look on Marty's face.

Marty avoided eye contact with Luc.

Looking ahead as if in a daze, Marty asked, "Are you alright?"

Luc replied, "Yeah, are you alright?" Then, Luc asked, "Have you seen me old man?"

Marty had a flashback. They were in the canteen and someone was calling out the emergency roll call. The names of workers on that shift were being called.

"Jim."

"Yes."

"Harry."

"Yes."

"Marty."

"Yes."

"Norm."

"Yes."

"Peter."

No answer.

"Has anyone seen Peter after the explosion?"

No answer.

"John."

No answer.

"Has anyone seen John after the explosion?"

No answer.

Marty turned to Luc and answered, "No, I haven't seen him, mate."

Luc gazed with begging eyes at Marty, "Was he on the plant?" he asked.

Marty couldn't think of the right words to say. "Look, I don't know, mate." Then, as if to distract Luc from the reality, Marty asked, "Are you right for a lift home?"

Luc looked at Marty and answered, "Um ... I'm not sure Marty."

"Mate, you better get home to your mum and you better be there with her," said Marty as he turned and walked away from Luc to conceal his tears for Luc's loss.

By this stage, Luc suspected that Marty knew his father may not have survived the accident. Luc's thoughts shifted to how he was going to tell his mother.

CHAPTER FIFTEEN

LOCKY WILSON

> WHAT WE HAVE ONCE ENJOYED WE CAN NEVER LOSE. ALL THAT WE LOVE DEEPLY BECOMES A PART OF US."
>
> *Helen Keller*

Locky was outside the house, washing windows. She heard the home phone ring. Mildly annoyed at the interruption to her chores, she put down the cloth and window cleaner spray bottle she was wiping the windows with and walked into the kitchen where the phone was. Her friend's husband was not his jovial self. As soon as Locky answered the phone, he said, "Locky we heard there's been an explosion at the gas plant."

Locky replied, "Don't be bloody stupid! I can't see any smoke coming from the plant."

He then said, "Why don't you ring and find out?"

She replied, "If you are right, I can't ring the gas plant. Peter will come home and bloody rip shreds off. If there's been an accident, he'll be so busy trying to sort things out. I just know not to interrupt his work."

After the formalities of farewells, Locky placed the phone back on its cradle and continued to clean the windows. Having washed the windows and cleared the rest of the chores around the house, she decided to drive to the bank to do some banking. This was one of her routine responsibilities as Treasurer of the Sale Football Club.

As she drove down Montgomery Street, she noticed two men from the gas plant standing on the footpath and talking. She stopped the car and walked to them.

She looked around and asked, "What's going on?"

One of the men turned to her and said, "Locky, there's shit happening there. I was on shift and when I heard the explosion, I ran away and drove into Sale." Then he asked her what she had heard.

Locky shrugged her shoulders and said, "Whatever's on the radio. That there's trouble at the plant." Then, in disbelief she looked at the worker and asked, "Are you sure?"

The worker replied, "I'm telling you, it's happened."

Locky could not believe that the robust and dependable plant she and her husband were loyal to and loved had exploded. She could not even conceive the likelihood of such an event.

Then, as she looked around the street she noticed a car that was being driven by some men she didn't know stop nearby. Her son Luc got out of the car.

Luc walked up to his mother. She saw sorrow in his eyes.

"Mum, can you drive me home?" Luc asked timidly.

Locky was glad to see Luc because seeing him gave her confidence that everything was OK with Peter too.

As Locky drove off, Luc sat quietly in the front passenger seat, looking ahead and wondering how to start the conversation with his mother about her missing husband. He cleared his throat and took a deep breath.

Then he said in a hushed voice, "Mum, I don't have a very good feeling about this. They haven't been able to get hold of Dad. I'm not sure what's going on and I don't know what to do from here. No one's been able to get hold of him. I saw the explosion. I'm not sure exactly what's happened but when the safety people tried to contact Dad through the radio he wasn't answering."

Locky kept looking ahead intently, concentrating on her driving. Luc turned to his mother and looked at her. He wondered if her silence meant that she wasn't convinced that her husband of twenty-nine years was missing or that she refused to believe it because she could not face that possibility.

Eventually, Locky responded, "Ah, he'll be busy. He'll be busy. You know what he's like. He'll be running the plant today, so he'll be busy."

Luc realised that Locky's defences and gates of denial were up. He suspected she knew within herself something was wrong and she was buying time to work out what to do to support her sons.

When mother and son got home, Locky told Luc to call James and Brett while she busied herself. "James had heard the news on the radio and he called me before, but I didn't have anything to tell him," she said as she walked towards the laundry.

James was remorseful when Luc called him. He kept saying, "Dad left me a message on my phone this morning, but I thought I'd call him in the afternoon. What's the news, Luc?" Luc didn't know how to respond. James realised it was futile asking for information or news at the height of the confusion.

"I've packed my bags. I'll get Brett and we'll drive home tonight," James said before ending the call.

Luc sighed as he placed the phone down. "Poor Brett," he said under his breath. "I hope he'll be OK."

Back in Melbourne, James called one of his uncles to ask him to go with him to Burwood, where Brett lived. "I just don't want to tell him over the phone." James explained. "You know what he's like. He'll take it the hardest."

As Brett walked up to his house, he saw James and their uncle waiting for him. Immediately, he sensed something was wrong. James broke the silence. "Brett, there's been an explosion at the gas plant and they don't know where Dad is. I've come to pick you up to go home."

Brett was speechless. He looked at James and his uncle as he fought back the tears. He breathed several deep breaths and nodded.

The drive to Sale was quiet and seemed to take longer than the usual time. Brett looked for the Esso flare that normally illuminated the skyline. But this night, he didn't see it. This time, he felt like they were driving into the heart of darkness. This was an apt metaphor for what they were travelling into. The loss and the impact of that loss would soon feel real for the Wilson family.

As the hours passed and the news travelled, family and friends started to turn up to check on Locky and to see if there was any news on Peter. Locky walked back into the lounge room where everyone was waiting for news. She looked at Luc and asked, "Did you speak to James and Brett?"

Luc nodded and said, "Yeah, they said they'll drive down tonight."

Late into the night, James and Brett arrived. By this time, everyone knew something was wrong, but no one knew to what extent. The absence of any communication was unnerving. Luc turned to his brothers and quietly said, "Poor Mum's trying to hold it all together, but you can see she's rattled."

Eventually, at ten o'clock, news came by way of representatives from Esso. They said to Locky, "We can't find him. We don't know where he is. We know he was in that area. But we just can't find him."

That confirmed what Luc had been thinking: that his father was dead. He looked at his mother and he knew that up to that point, she had been certain that if her husband was down there he might be trapped under something and they would find him. Luc knew that his mother's default belief was that her lifelong friend and husband would be alright. Throughout the afternoon she had been saying, "He knows the place backwards."

Once the news arrived, Luc turned to his brothers and told them what he had seen and heard during the day at the plant. He said, "I don't think he's coming home, boys."

Immediately, Brett broke down and sobbed. Luc knew that Brett would struggle with this news for years to come.

James felt the emotional pain in his body. He was distraught, but he stayed silent. He saw it as his responsibility to stay focused and in control. Older children in families seem to have the DNA for remaining in control when their younger siblings struggle. James became rational and started thinking about his duty as the eldest, to support his mother and his younger brothers. He thought about the funeral arrangements. He recalled that only two weeks prior, his father had been asked to give the eulogy at the funeral of a work friend's wife, because of Peter's oratory skills. James decided he would honour his father by giving his eulogy.

Luc felt numb. He couldn't cry. He couldn't sense any emotions. He felt hollow.

CHAPTER SIXTEEN

ELIZABETH, SUE-ELLEN AND RHONDA HEAR THE NEWS

"A CHILL SWEPT THROUGH THE AIR, THE SORT OF GRAVEYARD KISS PROMISING BAD NEWS TO FOLLOW."

Katherine McIntyre, *Rising for Autumn*

The motor bike shop in Sale had its usual steady stream of customers come and go. One of those customers on this day was Elizabeth Ward with Haydn and Katlyn. The beauty of working in the schools was the ability to be home with the children during the school holidays, as was the case on this occasion. As they walked into the shop, they were greeted by the young male shop assistant.

"Afternoon, Mrs Ward," the shop assistant greeted her.

Elizabeth Ward smiled and announced her reason for being there. "Hi. My husband has an order for me to pick up today."

The shop assistant looked at Elizabeth quizzically and replied, "I don't think it's in yet Mrs Ward. Let me check with John" — John was the store owner.

The shop assistant picked up the phone and dialled a number. "John, is Mr Ward's order in?" he asked. The look on his face conveyed a response unrelated to the question. Finally, looking over to Elizabeth he replied, "Yes, yes ... Mrs Ward is here." Again, he looked over to Elizabeth. Awkwardly, he called out to her. "Mrs Ward, John wants to talk to you."

Puzzled, Elizabeth walked to the shop assistant who handed her the telephone. "Thank you," Elizabeth said to the shop assistant. Then into the speaker, "Hello, John."

John's voice was not his usual jovial tone. He took a deep breath and asked, "How are you Liz? How's Jim?"

Surprised by the question, Elizabeth wondered why today's errand for Jim was any different to other occasions. Why was John quizzing her? Obligingly she replied, "Uh ... good thanks, we're good thanks John."

Elizabeth was particularly perplexed by the next question John asked. "Did he go to work today?"

She looked around the store and hesitantly replied, "Yes ... yes, he did." Elizabeth felt uneasy.

After a long pause, John continued, slowly and caringly. "You ... you don't know, do you?"

With this question, Elizabeth knew something was wrong. Unintentionally she raised her voice in response, "No. Know what, John?"

She felt her heart sink when John answered. "I'm sorry Liz. There's been an explosion at the plant."

These words were followed by a tingling in her chest. She tried to retain her composure despite the shock. She looked around the store and saw Katlyn and Haydn staring at her.

John suggested, "Why don't you just hang up and call the company and ask about Jim?"

Elizabeth knew she and the children would get greater comfort from Jim's parents. She replied, "It's OK, John. I think I'd like to go to Jim's parents' place."

Unsure what to do with the phone, she held it out for a while and finally handed it to the shop assistant. She was shaken and feeling nervous. All the while, aware that her children were watching her, she tried to hide her shock and anxiety from them.

Katlyn's question denied Elizabeth the opportunity to think things through and decide on how to best break the news to them. "Mum, what happened?"

Elizabeth regained her composure and slowly walked over to the children. At first, she hesitated. She wanted to protect them from the awful reality. Eventually, she answered the question with the love of a mother wanting to protect her children, while knowing that sometimes there is no way of telling the truth and safeguarding them at the same time. She started slowly, "Um ... there's been an accident ... an explosion at the plant. And ... and ... we don't really know much at this stage."

Surprisingly, Haydn innocently responded, "Oh, we saw all those fire engines and ambulances and things everywhere in the street earlier, Mum."

Realising the children had not made the connection between the explosion and the chance that their father may be hurt or killed, Elizabeth took each child by the hand and led them out of store.

- - - - - -

Sue-Ellen started her day early. Dropping Hayden off at school meant lining up behind all the other people driving their children to the kiss- and-drop-off zone. She was hoping to get all her errands and chores done before picking him up in the afternoon.

After waving goodbye to Hayden, Sue-Ellen drove to the shopping strip in Sale. She parked her car, walked around to the back seat, where Sophie was buckled into her child restraint and she pecked her child on the forehead while she unbuckled her seat belt. Sophie reached for Sue- Ellen's face with her small hands and hugged it as hard as she could.

Sue-Ellen smiled. After unbuckling her daughter from the seat, she encouraged the child to slide off the chair. Sue-Ellen was highly organised and she knew exactly where they would start to finish in time for school pick-up.

Sophie skipped beside her mother, holding her hand. Sue-Ellen had completed her errands and decided to stop by a toy shop before going home. As she looked in the window of the toy shop displaying a variety of toys in all sorts of colours she heard a muffled thump in the distance. She stopped to focus on the sound, but she did not hear it again. She wondered what it was.

A short while later, Sue-Ellen heard smatterings of news and talk in town about an explosion at the gas plant.

Her heart sank as she thought of Marty and his work friends.

After picking up Hayden from school, Sue-Ellen drove home. On the way she passed countless ambulances and fire trucks. When she got home, she saw a big cloud of smoke hovering over the gas plant. She didn't know if her husband was safe.

Her heart heavy and her mind troubled, she tried to maintain a daily routine for the children, to avoid alarming them. As she prepared the after-school snacks, she felt the tension build up in her entire body. She just wanted to know if her husband and the other people on the shift were safe. But it would be hours before she received any news. The silence was agonising.

Rhonda Rawson was in the car driving home when she saw a big black cloud over the plant from a distance. She thought, Ronnie always says there's a big black cloud at work. It's probably a hard day.

Then there was a newsbreak on the radio, broadcasting the explosion and fire. The news also mentioned that there were a few people missing. She turned the car around and went back to her office in Maffra.

She thought that any emergency calls would come to her office. She tried calling Esso but could not get through. Ronnie's mother rang her and they could not make sense of what was happening. Everyone had some news but no details and nothing to find comfort in.

She waited anxiously for hours. She thought about the children and wondered how things would turn out. She desperately hoped her husband had survived.

Eventually, the phone rang. It was someone from the plant. Rhonda's heart dropped.

The man introduced himself and said, "Rhonda, I was at the plant and when we were evacuated to the meeting point, I went around and took names and numbers of families I could call on my mobile. I'm calling you to let you know Ronnie is OK."

Rhonda felt the release of the tension throughout her body. The lump in her throat made it hard for her to speak. She thanked the caller. She sighed a sigh of relief because she knew that as long as her Ronnie was alive and well, they would be okay.

CHAPTER SEVENTEEN

MORE EXPLOSIONS AT THE PLANT

"WE FIGHT IT DOWN, AND WE LIVE IT DOWN, OR WE BEAR IT BRAVELY WELL, BUT THE BEST MEN DIE OF A BROKEN HEART FOR THE THINGS THEY CANNOT TELL."

Henry Lawson, When I Was King and Other Verses

The events that were unfolding were an assault on the senses of the people on site. The stench of burning gas, oil and vapour mixed with the smell of burning flesh, the sight of a mangled body with smoke rising off it; the horrendously loud and piercing sounds all competing for pitch. The blaze was like a nuclear attack.

A mix of high-pitched alarms set off by the explosions combined with warning sirens and emergency vehicles. People calling for additional ambulances were joined by the noise of the fire horn. The loud thuds of particles hitting nearby vessels became deafening. The roar of escaping gas and liquid and the whooshing sound of gas igniting followed by enormous bangs shook a nearby building to its foundations. A huge fire enveloped the entire area surrounding two of the gas plants. Amid these were the groans and cries of the wounded.

A valve in one of the plants opened and steam rose. Then suddenly there was a "boom" and a violent release of thick, white vapour that resembled a fog cloud rolling out. This was followed by a long rumbling sound like a distant thunderstorm.

A breeze moved the cloud slowly. Suddenly there was a "woof" sound followed by fierce orange flames. The fire was fuelled by nearby gas pipes rupturing and exploding and getting worse. The persistent efforts by some of the workers to extinguish the fire by putting water monitors on it were futile.

One of the explosions produced a huge, swirling cloud of stones, dirt and gas. One after the other, the explosions in the plant continued; at least one released 2700 kilopascals of air pressure.

The once orderly plant had turned into a battlefield, as if the adversary had ambushed the camp, capturing the innocent and seizing the opportunity to overwhelm them with fear, confusion and terror.

Workers standing nearby were thrown into the air like rag dolls and dropped to the ground and pipe racks nearby. The freezing cold liquid spilt over their bodies and burned their flesh. Some lost consciousness, while others shrieked and yelped as they writhed from the excruciating pain, their flesh sizzled to the bone by the flames from the fire. The disgusting stench of burning flesh mixed with the smell of burning gas filled the atmosphere. It felt like a war zone.

Peter, thrown into the air, fell and fractured his skull. He was knocked unconscious and fell at the base of the fire that incinerated him.

Nearby, John was blasted away as if by powerful ammunition. His body dropped to the ground, mangled and in a heap. Every bone in his body was crushed. Lifeless, his head dropped to one side and blood started to stream out of the side of his mouth. The impact of the explosion killed him. The fire incinerated him too.

Throughout the plant grounds, the blue metal, in the form of small rocks used to cover the clay-based soil around the site to make the surface more stable to work on, was thrown into the air like machine gun bullets from the pressure of the explosion. The shrapnel pierced workers standing nearby.

At another end of the plant, a pilot light on a heater ignited the hydrocarbon vapour and the ruptured vessel caught fire. Workers nearby were incapacitated and burned to varying degrees, on top of the cold burns.

Adam was thrown through the air by the force of the explosion. He tried to get up but could only lay there, trying desperately to lift himself off the ground, freezing cold liquid spilling over his body.

He shrieked with the pain, intensified by the heat around him. He tried to scream as if to alleviate the agony, but his voice was muted by the pain. The only release he had was to moan in agony. Softly he prayed, one syllable at a time, "God ... don't let me die. I want to look after my wife and kids. My baby girl ... my kids ... Please, God. Please."

He tried to get up again and cried out with pain as he pressed his right leg. The hot blue rock etched into his burnt flesh, which sizzled. He shrieked. Panting for air, taking short and shallow breaths he pleaded in a dry coarse voice to anyone who could hear him, "Help ... someone ... please ... help me ..." His body was twisted and the flesh on his thigh scorched to the point where it resembled a cauliflower.

The explosions continued, tearing off the safety goggles and hard hats of workers nearby. The men were thrown to the ground, unable to comprehend what had happened and how to react. The fire burned their clothes and covered them in soot.

They appeared drunk and as they tried to stand, they were shaky at the knees. Some were bleeding from the lips, eyes, ears and nose. One of them had a huge cut across his forehead and the flap of skin hung down over his eye. Another was bleeding from the ears and mouth. They were all in shock and disorientated, having trouble communicating in any way.

Jim was still in the control room. Immediately, he shut in the pressure recorder and controller, PRC4.

He looked around to his left and looked at the one of the technicians who was in the control room. For a second, time stood still. Both men were stunned. They looked at each other with uncertainty.

Jim turned to his right and tried to see through the glass of the southern control room door. He saw a cloud of vapour travelling from east to west. He ran to the door that had been left ajar after Peter left the control room. He saw two people wobbling along the footpath. They were both burnt. They appeared drunk and were wobbly at the knees.

They seemed to have superficial cold burns; one of them had a significant cut over one of his eyes and the flap of skin was hanging down over his eye. This was most likely caused by shrapnel.

Jim ran behind the Gas Plant 1 control panel with another technician and they activated the fire alarm.

"Let's go to the south door. See what we can do!" Jim yelled.

As they ran to the south side of the control room, they saw the smoke and several fires. Two more explosions rocked the plant.

The technician yelled, "Shit! We can't do a bloody thing! Melbourne's gonna burn down. The bloody fire is going to travel through the gas pipes into people's homes."

Suddenly, there were more explosions. The walls rocked. One of the explosions caused gas to come into the control room. Unbeknown to the people, the plastic coating on all the electrical cables was melting and the fumes and vapor was coming into the control room. It was toxic.

The thought of the devastation propelled Jim to keep the city safe from what had just been described. Shielding his face from the heat he yelled, "I'm not gonna let that happen! We gotta activate emergency shutdown 1! That'll isolate the crude oil from offshore coming into the plant and to the other end!"

"There are no fucking emergency shutdowns in the control room!" said the technician.

Then, Jim ran back to the door and looked out at GP922. It was in flames that were impinging on the pipe work above the walkway. A breeze was blowing from the east.

Jim knew he had to activate the Emergency Shutdown 1.

He looked at the technician who was close by. This was a very dangerous situation for Jim. He had to run ten metres towards the gigantic fire. Neither spoke but both communicated non-verbally. The technician looked at Jim as if to say "you're mad" because Jim would have to run to the fire.

Jim looked back at the technician as if to say, "If I don't make it back, you've got to go!"

Both understood the gravity of the situation.

As Jim stood on the platform of the south end of the control room looking at Emergency Shutdown 1 and the fire some twenty metres past that, he had one thought. That was, in thirty years of operation of the gas plant, this switch had never been used in response to an emergency. It had never been activated in a real-life situation. Tests had been carried out on it to make sure it did what it was supposed to do, but it had never been activated. Everyone knew that the production of the gas depended on this one switch.

Jim thought, *crikey, I could get into a lot of strife by activating this. If this is all just a figment of my imagination and this fire goes away in five minutes, I could be in a lot of strife.*

So, Jim opened the door and ran down the path into heat and smoke and activated the emergency shutdown switch 1. He tried to activate the other switches too, but the heat and smoke were too much.

He ran back to the door and on the way, he noticed people under the pipe rack on Control Room Road. None of these people were injured or burnt and their overalls were intact. He re-entered the control room.

By this time, other people had entered the control room through the north door. Several injured workers had also been brought in to have first aid administered.

While Jim's heart pounded with the flow of adrenaline, his mind sharpened as he thought about who else he needed to contact regarding the emergency. He methodically notified people as he had been trained.

He initiated the emergency response callout by ringing the Hutchinson phone number that was in the control room. Hutchinson would methodically notify emergency services of the level of emergency and their responsibilities under that emergency protocol.

"We need a List One call-out!" Jim said.

The base of the explosion was about thirty metres from the escalating fire. Jim was surrounded with loud sounds coming from the wall of the flames: the thud of the explosions, sounds of the alarms, fire and people in the control room. These made it difficult for the operator and Jim to hear or understand each other.

"I'm sorry, I can't hear you. Please repeat," the female operator replied.

Jim repeated his request loud and clear. "We need a List One call-out!"

'I'm sorry. I can't hear you."

"We need a List One call-out!"

Eventually, the operator understood. Relieved that she had heard and understood him, he hung up the phone.

Immediately after, Jim dialled the guardhouse.

"There's been an explosion," he said. "There's been an explosion and fire! We need ambulances, fire trucks and a rescue vehicle to the control room as soon as possible!"

The guard responded in the affirmative and they both hung up.

Next, Jim rang the Long Island Point control room on a different handset. "We've had a fire at the plant and we've shut down the LPG flow," he said.

After this, he tried to ring the offshore control room in Gas Plant 2 and 3 control rooms. He picked up the phone, but the phone was dead. So, he picked up the radio and tried to call them on the radio, but he had no communications.

He grabbed the other phone handset and tried ringing the offshore control room operator on that line, but that line was dead too.

Jim turned to the technician and said, "We've got no communications!" The technician was busy treating casualties. Jim thought, *the lines have burnt down and I can't communicate via radio, telephone or plant telephone. This is going to make my job of letting people know about the emergency extremely difficult.*

At this point, Jim noticed that someone had put a fire extinguisher at the control room south door to hold it open. He went to the control room door and observed the fire. He saw that what looked like the GP922 exchanger was well and truly alight. There were explosions across the walkway from GP922. There was cladding peeling of the analyser huts. Insulation was falling out of the pipe rack. Flames were coming out of the top of the GP910 fans.

Jim was inhaling acrid fumes. His throat and lungs hurt from the fumes. It was difficult to breathe. He lifted the fire extinguisher and closed the south control room door.

There were acrid fumes in the control room and both the technician and Jim were worried about what they were breathing in. They went to the west control room door to grab breathing apparatus off the wall but there was none there. They must have been taken by someone else.

Marty ran into the control room. Suddenly, he couldn't breathe because of the fumes in there. He looked for the breathing apparatus and it was all gone. He tried to hold his breath as he ran past the panels. He saw workers stretched out on the floor, covered with blankets. Then he saw someone sitting in the chair, blood running down his face and wounds to his eyes.

Marty yelled, "Jim! Jim! I need a stretcher! I need a stretcher!"

Jim looked at Marty and asked, "Crikey! Where's the bloody stretcher?"

Marty placed his hand to his throat. He couldn't breathe. He started gasping for air. He asked Jim, "What's this shit?"

They both started to cough and wheeze as they lost breath. Jim replied, "It's hydro-carbon. Let's get out of here! The stretcher will be in the training room!"

They ran into the small office where trainees would sit and do their study every day. An orange stretcher hung on one of the walls. As they grabbed the stretcher Marty suddenly stopped. As if in a daze he said, "Shit! I was sitting next to that thing all day and didn't even remember seeing the bloody thing!"

Jim said, "Come on!" and grabbed the stretcher. "Who needs it? What's wrong?" Marty replied that someone was burnt.

They ran with the stretcher, out of the door of the control room. The supervisor was waving them over and pointing. Jim and Marty didn't talk. They couldn't hear each other over the noise. The supervisor continued to point, so they went around the back of the switchgear building, on the south end of the control room.

Jim saw the person lying in the stones. He was literally black from head to toe. There was steam coming off him like the steam off dry ice. He appeared to have no teeth. By now he was surrounded by a few other people.

Jim and Marty rushed over to him, threw the stretcher on the ground beside him and knelt. He kept saying, "Be careful of my leg. Be easy on my leg". Another employee arrived to help. As they picked the injured man up, bits of his overalls were coming off in their hands. Jim didn't recognise him. He thought he was an outside maintenance contractor. One of the other workers kept asking, "Who's that? Who's that?"

Then the injured man looked up at them and said, "It's Adam. It's me, Adam."

Jim thought to himself, *twenty minutes ago, I was talking to him. He was a fair-skinned, blond-haired, blue-eyed, handsome man, six feet two or three inches tall. Now he is a smouldering wreck.*

There was steam coming off his hair and his overalls were falling apart. Jim looked down at his leg. There was a compound fracture and there were bits of bone poking out through his skin. He was just generally burnt. Shattered. Jim thought, he won't live. Jim felt sick.

Another explosion shook the plant more vigorously and this time it brought high, scorching flames.

Marty turned to the others and yelled, "Let's get him out of here!"

Someone else yelled, "No, let's wait for the rescue people!"

Marty said out loud, "Screw it! There's six of us here. Let's get him out of here. What are the rescue people going to do that we can't?"

Adam looked up at Marty, nodded, smiled and pleaded. "Yeah ... please get me out of here, Marty." Marty looked down at Adam, whose mouth was a black cavity. Shaking his head in disbelief, he thought to himself, *where are his teeth?*

When enough people arrived to carry the stretcher, Jim ran back to the control room. Later, when Jim walked past the first aid room the doors were open to load the stretcher into the back of an ambulance. It was Adam. They'd stabilised him and they had a burn blanket on him and an intravenous drip. Jim thought, *he's still alive. I can't believe it. It's a shame because he'll die in hospital, it's such a shame.*

Jim went back to the control room to look for the supervisor. He passed him coming out. He was having difficulty breathing. He was gasping for air. There was another person in the control room. He had been administering first aid but now he couldn't breathe either. He had breathing apparatus next to him on the floor.

The control room, turned into a makeshift emergency room, was filled with horrified workers. As explosions continued, the windows rattled. The helpless injured workers screamed in terror, "We're gonna die! We're all gonna die!"

The situation worsened with yet more explosions that shook the room and shattered the glass in the windows. Seeing smoke start to fill the room made the workers more anxious. One of them started to yell, "We're gonna die! We're gonna burn in this bloody fire! Oh God, my wife and kids! Oh God!"

The rest of the workers looked at each other and their silence betrayed their terror.

In amongst the screaming and fear, the supervisor started to yell evacuation orders. "Everyone! Commence emergency evacuation! All! Commence emergency evacuation!"

Surprisingly, despite the commotion all around, the evacuation progressed efficiently. Workers helped the injured out.

Jim started to shut down the plant. His breathing was short and shallow; his mouth was dry. He looked up at the air conditioner which was blowing out hazardous gases and smoke from burning PVC conduit that carried electrical cabling into the room. He collapsed on one knee near the door of the control room.

He thought to himself, *am I going to die here?*

He pulled himself off the floor and ran out of the control room for air. Barely recovered, he came back to finish the shutdown. Another worker walked in and tried to help. He couldn't breathe so he sprinted out. Jim continued to shut down, running in and out of the room to gasp for air.

Then he went through the control room to the north door where there was breathing apparatus on the wall. He grabbed it and went outside. Someone from the fire crew helped him put the breathing apparatus on. He wasn't getting a proper seal on the mask because of his glasses. He went around the outside of the control room back to the west door.

Determined not to give up, Jim continued to speak to himself. "Right, I've got to go back in and make sure that we're shutting down the plant properly. I have to flare off gases. By gosh, I hope I'm doing the right thing." As he prepared to go back into the room, he heard someone shouting. "Jim! Get out of there! Get out!"

Jim yelled back in reply, "No! I'll stay here! I'm going to do the job I'm supposed to do!" He continued the shutdown, turning off switches, reading monitors, closing operations.

The breathing apparatus wasn't effective. Jim continued to have problems with the seal. He was helping an electrician who had come into the control room to check some electrical equipment like the air pressurisation units, blower fan control panel and so on. Occasionally, Jim lifted the mask to speak with the electrician. When he did this, he breathed more acrid vapour.

When the electrician moved on, Jim went back through the central control room. He opened the door and looked out and saw the fire had escalated. Then he saw some people trying to set up a ground monitor. He noticed firefighting foam drifting down the walkway.

He closed the door and walked briskly back into the control room. He walked behind the control panel and felt the bricks with his hand to see if the panel was going to be damaged by the heat. The bricks were hot and radiating heat. Jim thought that he was in danger if he stayed in the control room.

A few more explosions took place.

Before leaving the control room, Jim wanted to make sure no one was left behind. He walked through the operators' locker room, the toilets, the storeroom, the supervisors' office and the coffee room, yelling to see if anyone was there.

In the coffee room, Jim felt overwhelmingly thirsty. He tried to get water from the drink machine and the sink. There was none. Jim felt concerned that there might not be any water to put out the fire.

He left the control room through the north door. Another loud explosion shook the ground. He felt the heat on the back of his neck and legs. He was coughing a lot. By now, he had removed the breathing apparatus.

He went to the fire shed with the fire crew.

He and another worker thought they would see if they could help in the Emergency Response Procedure Room in the administration building seeing they could not stay in the control room.

While running through the canteen to get to the emergency response room, Jim quickly saw some of the injured workers starting to come back to reality. The shock not only set in but they also became extremely aware of their surroundings. They were aware that they were bleeding, that they'd lost their boots, their hard hats and their safety glasses were all blown off. They were aware that they were covered in black soot. Then they started to enter a second phase of shock and became extremely agitated and wanting to get something done about their injuries.

Jim and the other worker got to the emergency response room which was being run by clerical staff. Typically, there would be supervision of the clerical staff by a supervisor. However, on this day, the supervisors were all at the explosion site when the explosion took place. Two were killed and two were burnt. Ordinarily, these people would lead an emergency response situation.

The supervisor was on the phone with the manager in Melbourne who was obviously trying to get a picture of what was happening. He was asking questions such as "What's on fire?"

Jim answered the question as best as he could. The explosions kept erupting. Windows rattled and the building shook. People were frightened for their life.

"What about the headcount? There are people missing."

More through instinct than knowledge, Jim replied "Peter's dead." "How do you know that?" he was asked.

"Because he told me he was going straight to 922 minutes before the explosion," Jim replied.

Jim could tell by the look on the supervisor's face that he refused to believe someone was dead. When this information was relayed to the manager in Melbourne, the order was given to evacuate.

The fire of the Rich Oil Deethaniser tower escalated, causing the largest explosion up to that point. It triggered shock waves that were felt in the car park where people had congregated. They felt the heat of the fire from where they stood.

- - - - - -

Marty walked away from everyone else. He squatted down and looked around the site in a daze. Then he started to sob, cupping his face in his hands. There were three more explosions. As if devoid of any fight or flight response, he continued to wail and sob, loudly and inconsolably.

One of his workmates went to him and put his hand on Marty's shoulder, "You right, mate?" he asked. Marty was squatting, hunched over, cradling his face and head with his arms.

Marty kept crying, "No … no … no, I'm not. All I can think of is Adam, the poor bastard, and his wife and the new baby."

Marty put his arms down and slowly looked up. He was squinting and had stopped sobbing. He saw a man driving around in a 4WD with a foam tanker. In disbelief, he shook his head and asked in a low, inaudible voice, "What the heck are you gonna do with that bloody foam tanker? What the bloody heck are you gonna do with that foam tanker?"

One of the workers tapped Marty on the shoulder and said, "Come on, mate. Let's get you outta here." Marty stood up and was met by a paramedic who, along with the worker, walked him to the street outside the gates.

While the fires continued to burn, most of the people left the plant. Despite the sound of the fires and distant sirens, an eerie quiet descended over the plant.

Marty looked through the gate at the fires, wondering what had happened. His thoughts were broken when he heard footsteps behind him. Marty turned in the direction of the footsteps and saw Luc approach him. "Marty, have you seen me old man?" Luc asked.

CHAPTER EIGHTEEN

NEWS ABOUT JOHN AND PETER

"A GREAT MAN IS ONE WHO LEAVES
OTHERS AT A LOSS AFTER HE IS GONE."
Paul Valery

When Elizabeth, Haydn and Katlyn arrived at Jim's parents' house, Jim's father assured them that Jim was OK because someone had reported seeing him. Elizabeth asked if she could leave the children with them for a short while. She left and drove to the hilltop overlooking the plant. What she saw was unbelievable. That's when she comprehended the enormity and seriousness of the incident. The huge flames and smoke indicated to her that something catastrophic had happened.

When Elizabeth returned, Jim's mother was on the phone crying, holding the children close to her. She said it was the hospital and she was talking to the nurse. The nurse was quite reassuring and said Jim was in hospital for observation. Elizabeth was gripped by fear. She thought to herself, *I didn't think he would have been hurt.* She could hear Jim coughing in the background.

"Mum, I'm going to the hospital," she said as she ran to the car.

Haydn and Katlyn stood there motionless. Worried and confused.

Elizabeth made her way through the traffic leading to the hospital and walked briskly to casualty. From there, she was led to Jim.

As she followed the nurse, the realisation of the enormity of the accident overwhelmed her. She got more and more frightened by what she noticed and saw. Burnt overalls and ash covered most of the hospital floor. The overalls looked like they had been cut off the victims.

She was led straight to the room where Jim and another fellow workman were. That's when she finally faced the terror of the day. She saw Jim lying on the bed, a breathing mask secured to his face and over his head. She tried to look into his eyes but he avoided hers. She saw terror and distress on his face. He could hardly speak to her. He just kept shaking his head. He was crying and trying to pull the mask off his face.

Though the breathing mask muffled his voice, she heard him say, "Peter is dead. Peter is dead," while shaking his head and looking straight ahead. He repeated that phrase over and over again.

Elizabeth did not know who Peter was. She tried to comfort Jim and calm him down. But this failed. He kept repeating, "Peter's dead". She asked, "Who is Peter?" but he did not answer. He became very agitated. He just wanted to get out of the hospital. He finally removed his breathing mask and said, "I'm getting out." Elizabeth tried to stop him and told him she would find somebody to see him. She tried to calm him down. Finally, a doctor came to see him. He said to the doctor, "I got to go. I got to get out of here."

At Jim's insistence, he was discharged the night of the accident. Elizabeth drove to Jim's parents' home and picked up the children.

Robert was driving home from work on that fateful afternoon. He was listening to the daily news when he heard the broadcast about the explosion. His immediate thoughts went to his brother-in-law, John. Then he thought of his own wife, Anne, and his thirteen-year-old son, Matthew.

He drove faster than his usual peaceful drive to get home. He rushed into the house and ran to Anne. "There's been a terrible accident at the plant, love," he said.

Anne responded, "Oh my God! John!"

She ran to the phone and started to dial the numbers she had for Esso. As a former employee, she had built many friendships within the company. She tried over and over to speak with anyone to get news on her brother. It was as if she was immobilised until she could speak with someone. Finally, she was given some news. They hadn't conducted a headcount on who was there and who was missing.

This brought short-term relief to the family. Rob looked at Anne and nodded slowly. "There's hope, love. Maybe he's got away somehow—somewhere."

Anne wrung her hands. She was filled with mixed emotions: dread, fear and a glimmer of hope against all hope.

"Well, I better get dinner on the table," she said, more so because she needed to distract herself. The family ate quietly. It was getting late and there had not been any more news.

After dinner, the dishes washed and dried, Rob was clearing the kitchen bench while Anne sat by the phone waiting for someone to call her. Suddenly, there was a knock on the door. Rob walked to the front door of the house, turned on the porch light and opened the door. He was met by the Esso plant manager and a few other men.

Rob greeted the unexpected visitors and invited them in. He led them to the lounge room. The visitors explained some details about the events of the day and finally told Rob and Anne that John was one of the workers who had been killed in the blast. It was a terrible, terrible night. A horrific night. Rob was grateful for the unplanned visit from a neighbour's son who was playing with Matthew. He served as a great distraction so that Matthew wasn't exposed to the nitty-gritty, traumatic details of how his uncle had died.

Anne was totally devastated. She was upset and couldn't eat anything for a long time after the news.

Over the following weeks, her only comfort came from her long-time friend Sister Maree, a nun from the local Catholic Church. Anne wasn't a churchgoer but she loved to help out the church and Catholic school where Matthew went, whenever she could. She particularly loved the visits from Sister Maree. She tried to help Anne find some peace.

The hardest time for Anne came when she was asked to undergo sibling DNA testing to identify John. Because he was so badly burnt and because he did not have any teeth, he could not be identified through the traditional means.

Shannon had been at work all day. When her shift ended, she was relieved it was Friday. She looked forward to the weekend, especially the Friday evening catch up with her father and two daughters. She sat in the driver's seat and reached for the radio. As she turned it on, she remembered it wasn't working. "Oh shit. Quiet drive home then," she said to herself.

When she got home, the home phone was ringing. Her husband answered. He looked over to Shannon and stretched his arm, motioning to her with the receiver. "It's for you."

It was one of the neighbours of Shannon's aunt. Shannon wasn't impressed because they'd had some personal issues of late. Out of courtesy, Shannon walked to the kitchen and took the phone. Polite, but cold, she said, "Hi."

The neighbour said, "Shannon, there's been an accident at the plant." Shannon knew instinctively that this involved her father. The neighbour confirmed Shannon's gut feel, "Apparently your dad's missing. I'm sorry."

Shannon's heart sank and started pounding. Her legs grew weak. She felt herself fall to the ground. Unable to speak, she cried and cried inconsolably. She felt unable to deal with the situation. Then she started to bargain with herself. *Maybe he's OK*, she thought as the tears rolled down her face. *Just because he's missing, it doesn't mean he's dead.*

But in the recesses of her being, her father's words kept echoing in her head, "I won't be around to see that. I won't be around to see that."

Gradually, Shannon found the strength to stand up and regain her composure as she waited for news. She was conscious of Ashley coming into the kitchen and seeing her. Eventually, into the night, she was told that her father had died.

The next challenge for Shannon was what she feared most: telling Ashley. She decided not to hide the truth from the doting granddaughter, despite the pain it could cause her.

Shannon sat with Ashley and said, "Honey, Grandpa isn't with us anymore. He went to work and there was an accident. Grandpa was really badly hurt and he ... he died."

Shannon looked into her daughter's eyes. She realised a three-year-old doesn't understand the finality of death as an adult does.

The weeks and months that followed brought this realisation to the child. One day Ashley asked Shannon, "Mummy, if Grandpa has passed away ... well, where is he?"

Shannon did not have an answer because she too wondered where her father was. She replied, "Well honey, I don't know." She hesitated before repeating, "I don't know. He can be wherever you want him to be and he can be whatever you want him to be."

For the months that followed, Ashley would give an update to Shannon as to who and where her beloved Grandpa was. "I think Grandpa is a cloud today, Mummy," or "Today, he was a tree", or "a bird". Eventually, Ashley realised that no matter who or where her Grandpa was, he wasn't with her drinking milk and eating biscuits on Friday evenings.

Shannon was determined not to let John Lowery disappear from her children's memories. From that day, she started to collect newspaper clippings, court case papers, death notices and anything about her father, which she kept in a box hidden in the house. She promised herself that one day, when she had courage, she would start a scrapbook for her children so that when they got older they would be able to read for themselves and understand.

Kerry had now been living in Perth for some time. She had managed to hide how unwell she was from everyone back home in Sale. The depression and anorexia had resulted in a breakdown which had required extensive medical treatments. She had given up work to have time to recover and receive the treatments.

On the afternoon of Friday 25 September 1998, Kerry was in her car driving home. She felt restless and turned on the car radio for some easy listening music and unobtrusive company.

The radio program was interrupted abruptly. The sternness in the voice of the female announcer revealed ominous news, "There has been a massive explosion at the Esso treatment facility at Longford in south-eastern Victoria. The initial reports are sketchy. Sources have suggested there will be fatalities, given the enormity of the explosion. We will bring listeners updates as we receive them."

Kerry felt a knot in her gut. She had a premonition that this had something to do with her beloved brother John. Involuntarily and immediately, she felt the nausea rise. She stopped the car on the side of the road and vomited violently over and over again. It was as if her body was rejecting the insight she had about her brother being involved in the accident.

After a while, the vomiting subsided. Kerry sat back behind the wheel and positioned her head on the headrest of the driver's seat. She felt weak and her entire body trembled from anxiety.

Gradually, she regained enough strength to complete her journey home. Immediately, she called her sister Anne, who confirmed that there had been an accident but they had been told that John was OK.

Kerry felt a sense of relief. Still holding the phone, she sat in her recliner, sighed deeply and said, "Oh alright, that's good." She asked Anne to call her back with any news.

An hour later Anne called Kerry, who was still in the recliner recovering from the toll the vomiting had placed on her frail body.

"Kerry, we've just had word that they can't find John."

Kerry felt angry and frustrated. "Oh, far out, when will they know for sure?" she asked in desperation.

"They don't know," replied Anne.

Kerry pleaded with Anne, "You worked at Esso, can't you call someone you know? I feel so weak and worried."

"Kerry, I have called them and they're calling me. It's just so hard. There were so many people on site, ambulances and other emergency services. There's so much going on, even now." Again, Anne promised to call back with any new information.

The final call from Anne confirmed her worst fears. John had been found dead.

As Kerry tried to accept the final news of her dear brother, she recalled one of their last phone conversations during their weekly catch-up.

"Sis, something's wrong, something bad's gonna happen."

"Why is that, mate?"

"Ah, they're not bloody fixing things they need to fix. You know I got a fanatical memory about parts that need to be changed. They're not doing the maintenance upgrades, they're not changing parts when they're supposed to change them."

She wondered, if the company had done all the right things and changed the parts and kept up the maintenance, would John still be alive.

CHAPTER NINETEEN

TWO FUNERALS

> "NO ONE EVER TOLD ME THAT GRIEF
> FELT SO LIKE FEAR."
>
> C. S. Lewis

The funerals for Peter Wilson and John Lowery were held over two consecutive days at St Mary's Catholic Cathedral in Sale. The cathedral, built in 1886, was adorned with French-made bells for the free-standing belfry in the cathedral grounds, and statues of the Sacred Heart, the Madonna and hand-painted ceramic Stations of the Cross.

Kerry wasn't sure if Esso had paid for her plane fare and accommodation because they were good or because of a sense of guilt and obligation. Either way, she was grateful she didn't have to find the money to get back home for John's funeral.

Once she was back, the hard part was waiting for John's body to be released by the coroner so the family could start preparing his funeral. It was overwhelming. Kerry would often wonder if it was a bad dream, like the nightmares she and her siblings had often had since childhood.

Finally, when the family was told they could start preparing for the funeral, they didn't know where to start. As they sat around the kitchen table at Anne and Rob's home, they were joined by a member of John's fire brigade and someone from the local RSL, where John had been a member. These men helped organise the funeral. Rob also contacted the local funeral directors who had been their family friends for a long time. They too said they would take this burdensome task off the family.

Kerry looked around and when she saw the numbers of people coming and going and offering to help because of their friendship with John, she realised that he was not just her big brother, he was a friend and good companion to so many people. She felt overwhelmed by their love for him and numb from the pain John's tragic death had caused her. At the same time, she was thinking about what she would do next in her own life and how she would settle back in Sale. Rob and Anne assured her that their home was her home for as long as she needed it. She was deeply grateful for their unwavering love and support.

Typically, the day before a funeral, family members are usually given the chance to visit their loved one for the last time. The coffin is left open and family come to see the deceased after they have been prepared by a funeral home. John's family didn't get that opportunity. They were told they could visit the funeral home and visit the coffin but not John's body. Ashley had painted something special for her grandpa and Shannon wanted to leave a special photo with him, in the coffin.

However, when she walked into the room where the coffin lay, she stopped mid-way. She thought to herself, *Dad was just over six feet tall, big and solid. That coffin's too small. That's not Dad.* Then the shocking reality of being burned in a ferocious fire hit her; the effects a fire that fierce would have on the human body.

On the morning of the funeral, Rob, Anne and Matthew travelled together. Rob was in awe of the numbers of people who attended the funeral service out of respect for John.

As the family walked into the cathedral, Anne looked at Rob and said, "Look after Matthew for me, don't worry about me, I'm fine. I'm fine."

During the funeral service, Rob looked at the Romanesque onyx altar, the impressive stained-glass windows and large statue of Mary Help of Christians. The cathedral was filled with John's family and friends, who were joined by multitudes of current and former workers of Esso, members of the Country Fire Authority, mates from the Gippy, the RSL, and everywhere else John had obviously made friendships.

Kerry was there too. She felt overwhelmed and numb with the grief of losing her beloved brother. She felt like it was a bad dream, a nightmare. John Lowery was her big brother but she didn't realise he was loved by so many people. She thought to herself, *John may have been raised by a poor family, but he certainly got a royal rich farewell.*

When the funeral ended, everyone walked out. Rob looked at Anne and noticed she looked pale. Just as he was about to ask her if she was OK, she collapsed. Rob knew she was emotionally and physically exhausted. The ordeal had been too much for her. He stayed with her while everyone else went to the local fire brigade where the wake was being held. The fire trucks had been moved out of the way and everybody was invited to join the family for a cup of tea or coffee. Even though Anne wasn't well, she went to the wake for about five or ten minutes. Then she turned to the funeral director and asked, "Can you take me home?"

She was taken home and went to bed for a few hours of sleep. Later in the day, Sister Maree stopped by to comfort Anne and just sit by her side.

When Rob and Matthew got home, Rob felt drained and exhausted. He was relieved the funeral was over. He wondered how the family could cope with any more tragedies. With this heavy thought, he decided to spend some time in his garden. He walked to the front yard where he grew red and yellow roses, and a mix of other trees. Then he walked along the side of the house to the back yard where he had three vegetable gardens. He decided that next spring he would plant some carrots, tomatoes and cauliflower. He'd give the broccoli a miss, seeing it needed to wait until the earlier months of the year. With a deep sigh, he walked back to the house where Anne was with Sister Maree.

Luc was also at John's funeral. He saw the outpouring of the family and community grief and while he felt sympathy for John's family, he felt restless about the next day because he now knew the atmosphere he, his mother and brothers would be walking into the following day.

He was right. The large church was again filled to capacity for Peter Wilson's funeral. James took it upon himself to organise his father's funeral, write his eulogy and organise the pallbearers. Whenever he felt the tug at his heart calling him to grieve, he reminded himself that he needed to be strong for his mother, his younger brothers and, most of all, in honour of his father. He felt the responsibility to protect the family and take charge.

At the end of the service, James, Brett and Luc were joined by many of their friends, the young men Peter Wilson had influenced. As they carried the coffin on their shoulders and walked out of the church, they could hardly see through their tears. Suddenly, Luc felt a tug and firm embrace. One of his friends who was standing in one of the church pews reached out to Luc and gave him a firm bear hug. Luc was overcome with grief and sobbed uncontrollably. He kept saying, "I can't believe this is happening. I can't believe this is happening. I just can't believe it."

Then, he looked over to his mother and saw her distressed as he had never witnessed before. She had always been the strong one and she had carried the burdens of the family with courage and resilience. Today, she was a mess. She looked lonely despite all the people around her. Luc knew the intensity of her relationship with his father and he felt for her heartache. He walked away from the pallbearers towards his mother. He put his arm around her shoulders, held her tight and kissed the top of her head. "We'll be right, Mum," he said, wanting to reassure her of a future. "We'll be right. I'll make sure of it."

During the funeral, James knew that eventually he would need to contend with the grief he had buried. But that was a problem for another day.

That day came the second Christmas after Peter's death.

Up until that point, the family spoke about Peter often and the situation received a lot of attention in the community, the news and the media generally. Up to six months after Peter's death, James often saw Peter's and John's faces on the front pages of newspapers. He felt comforted lest the world forget about the tragedy. But when the attention stopped in the public domain, James faced another grieving stage. He thought everyone would forget his father and what had happened. He wanted the world to remember how his father and John Lowery had died so that it would not happen to someone else.

The second Christmas after the explosion, James packed his bag and walked to his car to go back home to Sale to join his mother and brothers. He placed his black duffle bag in the boot of his car and sat in the driver's seat. As he placed the car key in the ignition he started to sweat. His heart started to beat faster and his breathing slowed down. Unsure of what was happening to him, he felt a lump in his throat. He sensed that he was in agony, but he could not understand why. Then, he started to sob uncontrollably. He couldn't move. It was as if the tears that had been held back for the past fifteen months since his dear father's death had burst the dam.

He sat in the car for hours, crying for the loss of his father, his best mate; crying for his mother and for Luc. He and Brett had come back to Melbourne and escaped going back on weekends. But Locky and Luc had stayed in Sale. They faced the emptiness of the house every day and the Esso flare every night. They'd heard the talk in the community and all the hearsay every day for the past fifteen months. He cried for Luc, who was left dealing with their mother's grief as well as his own, while James was hiding away from it all, living his life in Melbourne.

Eventually, James gained enough strength to get out of the car and walk back to the house. He could not survive the drive to Sale and he could not face the family this Christmas. The psychological and emotional reality had caught up with him. His father, Peter Wilson, was dead.

Locky, Brett and Luc sat quietly at the Christmas dinner table. Locky felt that she had to stay strong for the boys. She sighed a deep sigh and said, "Well, I guess it is what it is. We'll let James deal with things his way. We know he loves us and we love him, even if he doesn't want to be here today because your father's not here."

Brett looked up at his brave mother. Tears welled in his eyes. Each time he tried to speak, the lump in his throat blocked his voice. Eventually, he took a deep breath and spoke. "Mum, it's gonna take us a long time to get over Dad's death. But if I was going to say something to Dad, I would say that he would be very proud of you. I remember one day not long after Dad's funeral, you got us three in the car and took us for a drive and you said, 'This is not going to wreck us and it's not going to ruin us. It's going to bring us closer together.'"

Then he choked up again. "Mum, you called us to unity. You gave us strength and you helped us pull together when we could have fallen apart."

Luc was crying and Locky looked down at the table, trying to be strong. "Mum, Dad would be proud of us all. And most of all, he would be proud of you because you called us to unity. And you know, I'm not happy with where I am in my life. But I promise you, Mum, I'm going to build me a future that you and Dad will be proud of. I'm going to find myself and settle and be so successful that if I could look Dad fair in the eyes, I would be proud of myself."

CHAPTER TWENTY

THE LONGFORD ROYAL COMMISSION

"THE HERO OF MY TALE... IS TRUTH."
Leo Tolstoy

Forty-eight days after the explosion, on Thursday 12 November 1998, the Longford Royal Commission held its preliminary hearing in the Supreme Court of Victoria.

The purpose of the Royal Commission was to investigate the reasons for the explosion and to make recommendations to prevent or lessen the risk of a similar catastrophe from happening again.

Because the Royal Commission had a special function, only those parties given leave or permission could be represented.

The council assisting the Commission stated the following with regards to the role of a Royal Commission:

"The function of a Royal Commission is fundamentally one of inquiry and investigation .... It has a limited lifespan, and following its report, ceases to exist. It seeks to discover the whole truth .... In addition, a Commission may serve a vital function of calming public anxiety and informing the community as to the circumstances of a major catastrophe," ... "It is not a court of law and although it can make adverse findings as to the conduct of individuals and corporations, it does not determine rights between parties or convict individuals or corporations of criminal offences. Its function is to make the fullest possible investigation of the relevant facts and make findings and recommendations in its report. ... Nobody appears before such a body except with the leave of the Commission."[2]

The organisations that were allowed to appear before the Commission included the Victorian WorkCover Authority; the unions, including the Australian Workers' Union, the Insurance Council of Australia, the Country Fire Authority, the Victorian

Trades Hall Council representing the unions of groups of people such as firefighters and inspectors; representatives of the families of John Lowery and Peter Wilson; the State of Victoria; two chemical companies; and representatives of members of the Victorian Parliament.

Jim sat in his recliner chair in the living room of the small cottage. He stared at the ceiling. And, like every night over the past year, he relived the horrors of the months leading up to the Longford Royal Commission.

Interview after interview, telling the same story over and over repeatedly. First to the police, then to WorkCover, the arson squad, management and the American lawyers. George was present for all of these meetings.

Then came the meeting that held him captive with rage and disgust. A meeting with the three American company representatives, George and himself, faithful George. Jim recalled him saying to the company lawyers, "The men have had a terrible time. They've faced enormous trauma on the day and since then they've had to show up for meetings and interviews. Some of them are quite intimidated by all of this."

The American company representative responding, "George, you can assure the boys they have nothing to worry about. They did nothing wrong. You fellows did a wonderful job. It was a runaway train and you had no control. Just assure the men they have nothing to worry about." The company representative glanced over at his fellow managers and smirked.

Then Jim had a flashback to the meeting with him, George, a number of lawyers and a lean oil expert, held in the chambers of the Queen's Counsel representing the company.

The interrogating lawyer looked at Jim, then, as if having pondered for long enough, turned to Jim and George, saying, "Thank you for coming in again gentleman. George, it's good to see you again. I was telling my wife the other day that I've been seeing more of you than her lately, with you accompanying the workers to these interviews."

The room is quiet. George laughs innocently and uncomfortably before replying, "Well, I'm just doing my job."

Jim recalled the lawyer turning to him and saying, "Mr Ward, last time we spoke we neglected to cover a few issues, so we'll do that today, if that's OK."

Jim nodding and looking at George for reassurance, saying, "Yes, it's OK," then a long silence.

The lawyer looking at Jim intently asking, "So, Mr Ward, you have done the same job, more or less the same job, for nineteen years?"

Jim had started to answer, "Yes, when I started ..."

He was interrupted by the lawyer who said, "Let's work closer to today's date, shall we?"

Jim was taken aback. He adjusted his posture in his seat and sat back. He was unsure of what to say.

At this point in the meeting, the lawyer had introduced the lean oil expert. "Jim, this is Mr Flanagan who is a lean oil expert and knows everything about lean oil there is to know. Mr Flanagan and I will be asking you to clarify some issues for us."

Jim had felt some tension but trusted the process would be fair and just. He recalled the barrage of questions asked by the lean oil expert and the Queen's Council, and in the background a solicitor taking copious notes.

It was as if Jim's brain had put everything on hold and recorded and replayed the events of those meetings like a movie.

"What time did you wake up on the day?"

"Were you tired?"

"What is your understanding of the lean oil process?"

"Can you describe the procedure for turning lean oil into gas?"

"Are you a man of routine or are you more erratic?"

"What time exactly would you say you decided to call it an emergency?"

"Jim, what were you doing at 10.15 on the morning of the explosion?"

Jim recalled his reply and what followed. He had said, "Look, I'm not sure of the exact time of each event, I was completing numerous tasks at once and I …"

Then, as if to dismiss his answer, the Queen's Counsel had stood up, walked to the window and looked outside. He had taken his reading glasses off and chewed on one arm of the glasses before he had asked Jim, "Do you mean to tell me, Mr Ward, that this is the single most important day of your life and you cannot recall what happened at 10.15 in the morning?"

Jim had felt uneasy. He had looked over to George. They had made eye contact, nervously.

Afterwards, Jim and George had walked out of the Queen's Council's office into the elevator. Jim had turned to George and asked, "What was that about?"

George had been silent for a while. Then he had responded to Jim. "I don't know, mate. They didn't say anything like that to anyone else. Matter of fact, they were quite friendly and cordial to the other boys."

Jim remembered the grief and agony of that moment in the lift with George. "Shit, George, I reckon they're looking for a scapegoat. I reckon they just want a scapegoat. They don't care about me or anyone else. My own company; my own employer that I've given nineteen years of service to, that I stood on the south door of that control room, wondering about the loss of production, the fact that they were going to lose money, and I ran towards the fire that could have killed me and left my kids orphans. For that company! And then they turn around and do this to me!"

As Jim lay in the recliner in the lounge room, he felt the same angst he had felt on that day in the lift.

Elizabeth walked over to him, knelt beside him and softly whispered, "Are you going to come to sleep in the bed tonight?"

Jim roused from the daze and the flashbacks. He realised that he was not in the meetings with the lawyers but in his living room with Elizabeth.

He looked at Elizabeth for a short while and replied, "No. I think I'll just sit here for a while longer."

Elizabeth looked away; tears filled her eyes as she said, "Darling, you've been sitting here staring into space for a year. How much longer is this going to go on for? How much longer will we live like this?"

Jim sighed as he struggled for an answer before he replied with the standard, "I don't know, Liz. I don't know."

Elizabeth stood up and left the room, quietly sobbing as she wondered what was going on in her husband's mind as he lay there night after night.

Jim reverted to staring at the ceiling. Parts of the days in the witness box of the Longford Royal Commission hearing flashed before him like selected scenes of a movie.

First reading his statement then describing in detail his movements on the day of the explosion, followed by a multitude of questions concerning the information in his statement and his actions on the 25 September 1998.

Throughout the course of the inquiry, Esso sought to point responsibility for the incident to the individual operators.

Jim sighed, a deep and grief-stricken sigh. He picked up the report of the Longford Royal Commission, scanned over it and slowly read sections of it. It contained the harrowing sequential details before, during and after the fire and explosion. It recounted events which had occurred years, months and weeks preceding, which the Commission attributed to the culmination of that fateful day. In reality, these could only be depicted through a horror movie.

It described the effects of the explosion graphically. Of one worker, Kennedy, it stated, "He was blasted into the air, struck a solid object with his head and hit the ground with liquid, dirt and stones pelting him at high velocity. He felt as if he were being shot at by a machine gun. Wherever he crawled he continued to be pelted. He smelt nothing because he held his breath. He had his eyes closed. He crawled into the clear and noticed two other blackened persons crawling towards the control room. Kennedy stood up and fell over.

Blood was flowing from an injured eye."[3]

In later chapters, it described the impact of the explosion as "the force of this initial jet of gas and liquids dug a hole in the ground which was later measured to be approximately 1.5 metres in diametre and 1 metre deep". [4]

The report also described the images that had been viewed by the Commission through video. "The first video recorded picture of the fire shows a red yellow flame with a height of between fifteen and twenty metres and a thick cloud of black smoke drifting in a south-easterly direction."[5] And, "A ball of flames approximately forty metres in width and at least seventy metres high erupted ..."[6] "As can be seen from the pictures, the flame heights were in excess of one-hundred metres high and fifty-five metres wide."[7]

Concerning the bodies of the dead workers, the Commission said, "The bodies of Wilson and Lowery were removed at 5.20 pm on Saturday, 26 September, by the CFA with police assistance. By this time, safe access to the bodies was possible. Although the last fire was not extinguished until late in the afternoon on Sunday, 27 September, it was felt that the adverse psychological impact which the presence of the bodies might have on Esso personnel, warranted their removal."[8]

Then, slowly he read out aloud, "The ultimate cause of the accident on 25 September was the failure of Esso to equip its employees with appropriate knowledge to deal with the events which occurred. Not only did Esso fail to impart that knowledge to its employees, but it failed to make the necessary information available in the form of appropriate operating procedures."[9]

He turned to the proceeding page, as he had done on many previous occasions, and continued to read, "Those who were operating gas plant one on 25 September 1998, did not have the knowledge of the dangers associated with loss of lean oil flow and did not take the steps necessary to avert those dangers. Nor did those charged with the supervision of the operations have the necessary knowledge, and the steps taken by them were inappropriate. The lack of knowledge on the part of the operators and supervisors was directly attributable to a deficiency in the initial or subsequent training.

Not only was their training inadequate, but there were no current operating procedures to guide them in dealing with the problem which they encountered on 25 September 1998."[10]

Jim broke down and sobbed audibly and uncontrollably. The justice that was delivered by the decision of the Longford Royal Commission could never eradicate the overwhelming indignation and terror of being blamed for the accident that caused the deaths of Peter and John.

That night, like every other night, passed without any sleep for Jim. The fear of falling asleep and having the nightmare of that fateful day was far greater than the exhaustion of the lack of sleep.

CHAPTER TWENTY-ONE

THE WARD STORY CONTINUES

> "IT'S PERFECTLY SAFE FOR US TO
> TOUCH BASE ON OLD MEMORIES
> AND JUST MOVE ON."
>
> *Katlyn McDonough (nee Ward)*

Life was never the same for anyone in Sale after the explosion, least of all the Ward household. The things that had made the Ward family full of joy and laughter stopped. They all stopped. Jim didn't find enjoyment in them anymore. He didn't ask the children about their magic moments. He became withdrawn. He made some attempts to rekindle his relationship with the family but they were robotic and distant. Katlyn in particular felt this disconnect. When he asked about her school or sports, she felt as if he was just going through the motions. His lack of emotion, empathy or any connection to what was happening in real life was confusing to her.

Her hero had turned into his own nemesis. His patience was replaced with anger. His resilience was replaced by fear. Katlyn constantly felt the need to be careful of what she said, where she went and how she related to Jim.

Gradually, and over some time, she found it safer to retreat and hide from everybody. It was easier to go to her bedroom and be quiet.

However, at times, she would go and sit with Haydn. He seemed to be stronger and more comforting. One day, she went to his room.

"Haydn, can I stay here for a short bit?" she asked.

Haydn nodded and whispered, "Just shut the door."

Katlyn quietly shut the door, walking over to sit next to Haydn at the end of the bed. She looked down and stayed still for a long time.

After some time of quiet, Katlyn said, "Haydn, who are all these people who keep coming and going to our house?"

Haydn shrugged his shoulders. "I dunno. Sometimes I hear Dad say they're from the union, the media and people to do with the court cases and that."

Katlyn's eyes suddenly grew larger as she remembered. "Haydn, I heard Dad talking to some people and he said you and me are going to get some money. Do you know what that's for?" Haydn shook his head. She continued, "Do you think the money will fix Dad up and we can go back to a normal family again?"

Katlyn tried hard to hide her sadness from her brother. "The other day, I saw a family and they seemed so different to us. They looked happy and ... and normal. They didn't look like us."

Haydn looked at his sister and reached out and put his hand on her shoulder to comfort her. He smiled at her and said, "It's OK Katy. We're normal too. It's just that Dad isn't well."

Tears streamed down Katlyn's face, her blonde hair shielding her eyes as she looked down at the bed. "I don't know any more Haydn." Then she burst into tears. "What I find really scary is when he gets really angry. I'm scared of him. Sometimes I think he's going to hurt us or Mum or even himself." Haydn tried to comfort his sister. He kept his small hand on her shoulder as she continued. "Sometimes I don't sleep all night. I'm so scared of him."

"Yeah, but he never hurt us Katy," Haydn said, reassuring her.

Katlyn was silent again. Then she asked, "Why do Mum and Dad fight all the time these days?"

Haydn sighed deeply. "I'm not sure. They think we don't hear them so don't say anything," he cautioned.

Katlyn shook her head. "I won't say anything," she promised.

As the children sat and talked, Elizabeth came into the room. "What are you two up to?" she asked, masking her own anguish behind her loving smile.

Katlyn answered, "We're just talking about how weird things are at home, Mum," she confessed.

"What do you mean?" Elizabeth asked.

The children were quiet. Then Haydn replied, "Well, Dad's always angry and he yells at me for slamming the back door and everything. And there's always people here. We don't like it!"

Elizabeth tried to comfort them again as she had done so many times before. "You know Dad's not well. It's OK. If we just be patient with him and try to be quiet when we're around him, everything will be OK."

Deep inside, Elizabeth knew that the children were right. She too felt distant from Jim. There didn't seem to be any hope for the situation that had been thrown at them. Even a recent family holiday to Queensland had felt like hell to her. Jim had played golf, watched sport on TV and locked himself away. The children had found this very hard. She had found it harder.

On that fateful day in September 1998, she had lost her husband and at present she was living with someone she didn't know. She felt pain in her chest as she realised that the Jim she had known may never return. Her greatest fear was that he would hurt himself or take his own life. Finally, on one occasion when she had spoken to the counsellor, she had said that she was worried Jim might suicide. The counsellor had replied, "You have good reason to be worried." The next day, before going to work, she called his father and asked him to go over and stay with Jim.

She was scared of the unknown, and she was scared for Jim, especially when he attended the funerals for Peter and John. Elizabeth knew Jim had only attended a few funerals throughout his life and she was scared he didn't know how to cope.

One night, as Jim sat in his recliner watching Katlyn quietly drawing in her book, he was shocked when his daughter looked over at him and started to yell at him, angrily abusing him.

"You promised! You promised! You didn't keep your promise! You lied to me!"

Jim snapped out of his trance. He walked over to Katlyn and held her tightly to his chest. Jim asked, "What did I promise, baby?"

Katlyn pushed him away, replying with rage in her voice, "You promised that your place would never blow up!" Jim knelt in front of Katlyn. With tears in his eyes, he glanced over to Elizabeth, who was also crying.

Jim looked back at Katlyn and said, "Yep. I did promise. I made a mistake and it could happen again."

Haydn slowly walked up to Katlyn and Jim. He looked at Jim, and pleading with him, he said, "Dad, please don't go back to work. We don't want you to go back to work. Promise you won't."

Jim looked at them and said, "I'm sorry guys. I have to go back. This time I won't make a promise I can't keep."

The harder the Wards tried to return to some form of normal, the further it got away from them.

In the months following the decision of the Royal Commission, Jim found it impossible to work for Esso any longer. He couldn't tolerate working for an employer that was prepared to blame him and other workers for what happened on the day. Once it was proved that none of them was at fault and the company still refused to apologise, he just could not go back to the plant.

In addition, he was in the grip of trauma from the day and the deaths of Peter and John. So, he told Elizabeth that he was moving to Melbourne to find work. Elizabeth felt unable to influence this decision because they had all been living on tenterhooks since the changes in Jim.

He moved to Melbourne and took employment in the financial services sector to be trained as a mortgage broker and a financial planner. The following weeks and months placed a much greater strain on the family than anyone could have imagined.

Jim's income had been reduced considerably; the family expenses had been increased with the additional household to maintain in Melbourne. Jim became exhausted travelling to Sale every weekend while working weekdays and studying at night. Elizabeth felt torn between the children, who didn't want to be with him, and her unshakeable commitment to their marriage and wanting to make it work.

There were discussions about the family moving to Melbourne, or up the coast to NSW or to just get away to somewhere completely different. But these were illusions concealing the reality of the aftermath of the explosion and the impact it had on the family.

Finally, one weekend Jim came home and said that the marriage was over. He said that he still loved the family, but he needed to be in Melbourne. Elizabeth felt sick. She couldn't believe that despite the family unity and support over the recent past, it was ending this way. She wondered how she would cope with the children, as they dealt with their anger and typical teenage antics. The things she feared most were the shame, grief and embarrassment in a small community like Sale.

As tears streamed down her face, she wondered who she could run to for comfort and strength. She remembered that her father and brother were in a nearby workshop. She walked out of the house and drove to them.

When her father saw her face, he knew she was distressed and that she had been crying. After hugging her and trying to comfort her, he called her mother to come to the workshop and help console her. Eventually, after some time with her family, she returned home. Another one of life's battles was looming and she had to keep going.

In the early hours of the morning, some months after the announcement, Elizabeth was lying in bed. She knew she needed to get out of bed and take the children to school but she didn't have the strength. All she had energy for was to curl up in bed. She tried desperately to straighten up and get out of bed, without success. She picked up the phone and called her mother.

"Hello?" her mother's instinct sensed the gravity of the early morning call.

"Mum, can you please come and take the kids to school?" Elizabeth asked. "I can't get out of bed. I don't know what's going on."

Her parents arrived soon after the call. Her mum took charge of the morning routine, getting the children ready for the day and driving them to school. Her father quietly and patiently sat next to her on her bed and listened to her tell of her grief and loss.

"Dad, I'm so ashamed," she said, and wept uncontrollably. "I'm so embarrassed. I'm so tired and exhausted from keeping up appearances and protecting Jim and the kids. I can't do it anymore! I can't do it!" She stopped for breath before continuing. "I go grocery shopping either very early in the morning or really late at night, so I don't run into people I know. Do you know how tiring that is?" She kept describing her anguish. "Haydn and Katlyn are suffering too. I want to do the best I can as a single mum. At times, both reject their father. They won't speak to him and they won't answer his phone calls. He thinks I'm stopping them from talking to him. Dad, I want this nightmare to stop! I feel anxious all the time. I'm depressed. We all are, including Jim! I know we are! I hate this whole thing that's happened to us. Our family has been destroyed. My marriage has been destroyed. I want my husband, before the explosion, back!"

She wept, taking short breaths between statements. "The other day some of Jim's workmates were with their wives in the main street and when their wives saw me, they crossed the road and went across the street Dad! I know that was probably because they didn't know what to say to me. But it was so humiliating. When Jim was around, my life revolved around his shifts, day, afternoon and night. My whole social life and ours as a couple revolved around his shifts and around his work friends. When the explosion happened, all his work friends had their own issues to deal with. Now I'm left alone with the kids and no one to talk to. Some of the managers' kids go to my school. They don't even show they care. No one called from management just to see how I'm going! The most I get from the Esso parents at school is a cursory nod."

As Elizabeth sobbed, she continued to speak through her tears. "The other day, Katlyn said to me, 'Mum, I wish Dad had died in the explosion. Then I would have good memories of the loving and caring dad. I was his special little girl.' She's done that a couple of times. The accident killed the memories of their childhood with their father! All I could think about was the families who did lose their fathers in the accident. But I didn't know what to say to her.

"Dad, I don't want to demonise Jim! God knows he's suffered! His depression and post-traumatic stress after the blast, they changed him from being the warm, loving and supportive husband I married into a withdrawn, disturbed and troubled man. I don't even know him anymore."

Finally, her father reached over to her, gently holding her by the arms, he urged her to sit up. When she had gained the strength, she sat up. Her father put his arms around her shoulders and hugged her.

Elizabeth's mother returned from the school run and sat quietly on the bed next to her husband.

"Now you listen to me, Elizabeth," he said, his strong and commanding voice betraying his deep pain for his dear daughter. "You might not have Jim, your marriage or all the friends you used to have before the explosion." He stroked her hair, trying desperately not to cry with his beautiful and emotionally fragile girl. He continued, "You just remember you have me, your mother, your brother Peter, your sister Leonie, their families and your friend Pam." Elizabeth sobbed quietly.

Then, remembering the love and comfort she had received from these people, she looked up at her father and smiled sadly. Her father continued in a low gentle voice. "How many times have you told me Pam's called you to remind you to eat, or to look after yourself, or to take time out?" Elizabeth closed her eyes and tears streamed down her face. Her father gently took a tissue and patted her cheeks to dry the tears. He recalled having done so many times in the past when Elizabeth was a little girl and would bring her burdens to him. His heart ached, and he felt the weight of absolute and unconditional love for his dear child, who had been through so much trauma. He prayed that she would not break under its weight.

"I want you to promise me that you'll remind yourself of this every day! You have people that care about you and care about what happens to you. We will always be here for you. Do you understand?"

Slowly, Elizabeth felt reassured and somewhat stronger through her father's hugs and wisdom. "As for the shame and embarrassment, let me tell you. You have done nothing wrong. We're not ashamed of you! We are proud of you! You have stayed strong during a horrendous ordeal that would break others. You are caring for teenagers who have had to cope with impossible trauma. And you're doing a fine job. Do you hear me?" He looked intently into Elizabeth's eyes. "Promise me that you will walk strong and tall when you go out. Remember, I'm proud of you and your mum's proud of you. Your whole family is behind you! And that's all that matters!"

The quiet that ensued allowed Elizabeth time to think about what her father had said. "Now, I'll go to the kitchen and make you a cup of tea. You get ready, so your mother can take you to the doctor."

Elizabeth looked at her mother and felt comforted. She remembered the many occasions during difficult times when her mother's mere presence brought her the comfort she needed.

Meanwhile, Jim had kept running away from Esso. He had packed as much activity into his life as his adrenalin gave him energy to. Full-time work, part-time study, setting up a new apartment and furnishing it in Melbourne, driving back to Sale on Friday nights and trying to fit back into the children's routines. Friday night swimming lessons, dinner with family or relatives, Saturday morning kids sport and more family time on Sundays before driving back to Melbourne by ten thirty for a five o'clock start the next morning, was taking its toll.

After three years of trying to reinvent himself in the unfamiliar world of finance, he decided that his experiences, skills and passion were better used in promoting health and safety in workplaces.

Many opportunities were presented to Jim. He found himself and his world changing and growing. He found new insights into unions, regulators and the politics of safety and work. This led him to study for and obtain formal qualifications in workplace health and safety. Initially a Certificate IV in Work Health & Safety (WHS) and then a Graduate Certificate in WHS and finally a Graduate Diploma in health and safety management.

He found many and varied opportunities to talk with people like himself, about what they did for their living and what they were concerned about. He tried to deliver some sort of effective method to them for improving the opportunity to have a safer workplace. In the end, however, he grew disillusioned with a system that failed to place the safety of workers at the centre of its decisions. He gained insight into the way governments and organisations associated with the politics of work and safety used health and safety as a political football and, at times, for their own benefit. He felt disillusioned that the ultimate sacrifice of such a system saw workers getting shafted every day in Australia.

He was outraged that while safety laws had been in place for decades, for the most part they had been ineffective in securing and protecting the health and safety of workers. All the while, progressive governments gradually watered down the fundamental human rights of workers to take common law action against businesses and corporations not contributing to safety by not complying with the minimum legislation.

Jim's dreams and aspirations to change the world of safety were inhibited by the legacy of anxiety and diagnosis of Post-traumatic Stress Disorder after the explosion. This was exacerbated by the bitter experience of having his employer stare him in the eye and suggest that he was responsible for the events of the day at Esso Longford on 25 September 1998.

As the years passed, Elizabeth, Haydn and Katlyn started to adjust to their new life, each having to redefine their role in the family.

Elizabeth was astounded at the way Haydn instantly became the man of the house. He took on some of the chores and responsibilities he knew she struggled with. He would often help with the firewood, fix breakages and other problems that happened around the house. She knew he had by necessity grown up very quickly and ahead of his peers, and that made her very proud of him.

One Saturday morning, Elizabeth's parents stopped by for a visit. As they sat around the kitchen table, Elizabeth served them fresh sponge cake and coffee.

Her mother looked at her affectionately and said, "Liz, your father and I want you to know we are very proud of you and the children." Elizabeth looked at them with curiosity and smiled.

Her mother continued, "We were talking about all the things you have been through and how you have overcome so many obstacles and rebuilt your life. Then we got talking and said if we compared all you've achieved despite what you've been through, with other people who haven't had the same issues, then you, Haydn and Katlyn are champions. You have stood your ground in really tough times and won."

Elizabeth choked up as she realised the truth in what her mother was saying. Her father reached out and placed his hand on Elizabeth's. He smiled and nodded.

Finally, she said, "You know, Mum, I'd never stopped to think of it that way. I realise now that we have been strong because we have rebuilt our lives and moved on. The past will always be there, but we haven't stayed in it."

Elizabeth's face lit up and she went into deep thought. She nodded and said, "You're right. I think we were all feeling overwhelmed with everything in the beginning, but now when I look back and I see that over the years, little by little, one challenge at a time, we have morphed into stronger people."

She continued, "I remember when Jim and I first split up, I bought the house and took on the mortgage just so that the children had stability. I worked full time and was a full-time single mother. Even though money was extremely tight, we made ends meet. The bills were paid, and the kids used to joke and say I always had a contingency account to cover the cost of a new hot water service in case it blew up." Elizabeth's parents laughed with her.

Elizabeth sighed, "I'm telling you, it was hard work. Running the kids to sport and after school jobs, throwing Haydn's bike in the boot when it was raining so he could finish his delivery job, waiting in the car park of KFC at 11 pm for Katlyn to finish the late shift."

Her father nodded and said with pride, "Not to mention your high standards of keeping your home neat and tidy, just like your mother."

Elizabeth nodded and said sheepishly, "Yes. I do like to keep a neat and tidy home."

Her mother added, "I remember coming to see you on so many occasions and if you weren't ironing the children's school clothes, you were learning how to use the ride-on mower."

Elizabeth laughed as she said, "Yes. Haydn's mates always made jokes about me always being on the ride-on. Little did they know it was my time out and relaxation."

As the conversation continued, Elizabeth looked out of the kitchen window. She saw a monarch butterfly fly from flower to flower, its beautiful orange, black and white wings fluttering. She smiled and thought, *just like that butterfly, I struggled and fought until I broke out of my cocoon and transformed into who I am now*. Then she turned her attention back to her parents and said, "Thanks for coming today. I just realised I must tell Haydn and Katlyn how proud of them I am as well. Sometimes, it's so easy to get caught up in the pain that you don't realise you have broken through it and you don't remember to acknowledge yourself and those around you."

EPILOGUE

WHERE ARE THEY NOW?

WARD FAMILY

Jim and Elizabeth divorced in 2000.

Jim Ward obtained several formal tertiary qualifications in the Work Health & Safety (WHS) field, including a Certificate IV, a graduate certificate in WHS and a postgraduate program Graduate Diploma in Health & Safety management. He has remarried, and he and his wife reside in Melbourne. Jim retired in December 2017 because of his inability to work due to the ongoing anxiety from the events of Longford, including the subsequent attempt by Esso to suggest he was responsible for the events of the day. He was diagnosed with Post-traumatic Stress Disorder after Longford and spent ten years working with professionals on his mental health. He says that he will never ever psychologically be the person he was prior to the 25th September 1998. "While I'm largely repaired, I have a legacy and that legacy impacted on my capacity to do what I wanted to do in workplace health and safety in the union movement."

Elizabeth Coleman Gray (formerly Ward) continues to teach and is Deputy Principal at the local Catholic School in Sale. She is married to Jeremy, whom she refers to as her "Greatest support. He has helped me heal. He is my strength, my rock."

Elizabeth's message to anyone who reads this book is "Be strong because in the end that will guide you through. Try to have a voice. I regret that I couldn't have a voice in those days. But we were placed in a situation where Jim was blamed for the explosion and I didn't feel that I could have a voice. I had to be quiet. I didn't want to jeopardise proceedings for Jim. My message is stand up to work environments! Be brave and love your loved ones."

Haydn Ward lives in Maffra and works as a fitter and hydraulic technician in the oil and gas industry. During his time off Haydn enjoys spending time with his children. He loves playing guitar and making plans for his future permaculture farm.

Haydn's message is: "Money means nothing at the end of the day. Neither the compensation nor all the money you get paid working in the dangerous work means anything. If something goes wrong, it doesn't mean anything."

Katlyn McDonough (nee Ward) In relation to interviewing for this book, Katlyn said, "Mum was worried about me and Haydn to do these interviews; in particular about me feeling upset. So, I said to her we can do this. We do this for the sake of hopefully helping some other family who might have been affected by any sort of safety issues at work. And I said to her it's perfectly safe for us to touch base on old memories and just move on. It's not going to harm us. They're just memories. So, we had a talk this morning and she felt a bit better after that. I said to Mum, from what I gather she would have been very reserved and calculated as to what she said to anyone at the time. It was to protect me and Haydn and Dad as well. I said to her, you're divorced from Dad now and we're grown adults and it's your turn to speak up. She doesn't have to protect us anymore. Nothing she says about what happened during those times we don't know already. She might think we don't know but we were behind the lounge room door listening to the conversations and arguments. We heard everything. I said to her speak from the heart. It's her turn to speak the truth. There's a lot that can be learned from her perspective. She should feel comfortable with speaking openly."

Katlyn's message is to value human life. Employees are not just one human. They come with a family and extended family and lives behind the closed doors. What happens to them at work has such a large ripple effect at home.

Katlyn is happily married to Aaron and recently they had their first child, Banjo, both of whom she is immensely proud.

WILSON FAMILY

Locky Wilson lives in Sale. She has been coaching basketball since 1980 when Jamie started playing and continues to coach her grandchildren. Locky was on the Sale Elderly Citizens Village board for six years until she resigned in 2014. She is a volunteer with Meals on Wheels and with the Catholic Church bereavement group for catering with funerals.

James Wilson lives in Perth with his wife Bianca and their two children Sonny and Meela. A qualified secondary school teacher, James changed careers soon after his father's death and is now a full-time musician. James continues to be amazed at the strength and passion displayed by his mother Locky over the past twenty years. She has been the rock which has kept their family together despite her own personal pain. It is her influence that continues to inspire him to be a better father, husband and family member. He also takes much comfort in the thought that wherever his father is, he will be getting much enjoyment out of the fact that his two youngest children have less hair than he ever did.

Brett Wilson is finishing his third university degree and is Assistant Principal at a Catholic School in Adelaide, where he teaches and is Head of Religious Studies. Brett and his wife have three children. Brett's deepest desire is that the world does not forget the tragedy at Esso Longford. He keeps his father's memory alive in his children by talking about him to them. They often say, "When I die I'll go up and be with Pag," which was the nick name Peter's sons had given to him when he was alive. Brett believes that the hope to be united with the grandfather they did not meet in this life has helped his children not fear death because they know he will be up there. Brett says, "When Dad died, I wasn't happy with where I was in my life. And now I could look him fair in the eyes and be proud of myself."

Luc Wilson eventually sought professional help because he wasn't coping with the stress and anxiety after the experience of the explosion. After the birth of one of their children, while Luc worked offshore for Esso, he and his wife decided to return to Sale where Luc would apply for work in the plant as an operator. One of the processes in the recruitment of plant operators is for prospective employees to tour the plant. One week before the tour,

Luc started having nightmares about working there. Unable to process the intense emotions, Luc started drinking alcohol to quash the anxieties and he came home intermittently. At this point, his wife gave him an ultimatum to straighten himself out or she would leave and take the children with her because she didn't deserve to be treated like that and neither did the children. Luc sought professional counselling. He says it was the best thing he has ever done. He encourages anyone who is facing emotional and mental ill health to get help and talk. Luc lives in Sale with his wife and three children. He has been working for Esso for the past twelve years.

JOHN LOWERY'S FAMILY

Kerry-Lyn Walker lives in Adelaide with her two little rescue dogs. She works for a crane company as an office assistant. She has a daughter, Danielle, who has given her four grandchildren that Kerry loves with all her heart.

Kerry has two wishes. First, that John and Peter be remembered with respect. And second, that Esso make sure they don't let this disaster happen again so that no one ever goes through what Kerry and the rest of the family lived with.

Robert and Anne Bumpstead In 2014, Anne died of health complications after receiving a lung transplant. Robert Bumpstead is retired and he and his son, Matthew, live in Sale where they enjoy spending time together and occasionally go on holidays overseas.

Shannon Lyons lives in Sale. She has a collection of information relating to the explosion and her father's death, including newspaper articles, death notices, court case records and magazine clippings. When she has the courage, she will make a scrapbook of these items to give to her children to make sure the life of John Lowery is not forgotten.

JACKSON FAMILY

Marty and Sue-Ellen live in Longford near Sale. They have three children Hayden, Sophie and Jessica. Marty continues to work for Esso.

Following the incident, Marty started having graphic nightmares, which included seeing the face of his burnt workmate staring at him and his children being engulfed in flames and burning to death. He experienced fluctuating emotions and on a number of occasions would wake at night in a cold sweat, sitting upright in bed crying. He became hypervigilant and certain smells and noises triggered vivid memories of the day.

Marty was diagnosed with Post-traumatic Stress Disorder for which he sought professional help. His son Hayden began doing what Marty explains as "strange drawings" and as a result they also sought professional help for him. Sue-Ellen would regularly drive Hayden to the gas plant when Marty was at work to reassure him that everything was alright.

Marty spends a considerable amount of time at his beach house located on Ninety Mile Beach, where his family is actively involved in the local surf club.

Marty would like to acknowledge his wife Sue-Ellen for being a pillar of strength through some challenging times and also those Esso employees and community members who displayed outstanding courage during this time of adversity. These are the unsung heroes not mentioned in this book.

Honourable Bill Shorten MP

Member for Maribyrnong, Victoria

Bill Shorten served as the Secretary of the Victorian Branch of the Australian Workers Union until August 2007. His time as secretary was marked by a reform of the union's structures and expansion of union membership. In 2001 and 2005, he was elected as the union's national secretary. In 2007, he resigned as Victorian State Secretary of the AWU. Since then, Mr Shorten has held several prominent positions within the Australian Labor Party, the executive of the Australian Council of Trade Unions and the Superannuation Trust of Australia (now Australian Super) and the Victorian Funds Management Corporation.

Another high-profile workplace tragedy Mr. Shorten was heavily involved with was the Beaconsfield Gold Mine collapse when, in his role as National Secretary of the AWU, he was a negotiator and commentator on developments in the immediate aftermath and the ensuing rescue operations.

In 2007, Mr. Shorten was elected to the House of Representatives as the Labor MP for Maribyrnong. Since 2013, Mr. Shorten has led the Australian Labor Party as the Leader of the Opposition.

Ronald (Ronnie) Rawson spent the next two years after the explosion helping Esso rebuild the plant in the role of Permit Supervisor. Following this, he had the opportunity to work on and manage a number of large-scale projects overseas in the United States (Houston), South vKorea and Russia (on Sakhalin Island).

He attributes his ability to heal and regain normality of life after the explosion to the unfailing support and love of his wife, Rhonda.

Ronnie retired in 2010 and he and Rhonda currently reside in a lifestyle village in Ballina, NSW.

Robert (Rob) Miller was affected emotionally and mentally. He had recurring bad dreams and nightmares and suffered from memory loss. His marriage broke down soon after the explosion. He is happily married and lives in Cairns and works for Exxon in Papua New Guinea as an operator and, as part of this role, he mentors international employees. He still has his roots in Sale and returns home to visit his children and family and friends when he has extended leave. His message to workers is be honest with yourself. Don't get trapped into taking shortcuts or ignoring safety rules.

George Parker was awarded life membership by the Australian Workers Union, Victorian Branch in recognition of his service to the union members at Esso Longford. He continued to work at Esso until his early retirement due to ill health. He died in his home in 2015.

Of George Parker, the Hon. Bill Shorten said, "I don't know if the men realise George Parker was the true strength of the union effort at Longford. He was a special man. I miss him. He was the ideal shop steward. The best worker in the plant and the most respected worker in the plant and the union delegate. I wish he was around to see what I've done now."

HOW IT ALL ENDS

This book ends with the future in mind. I encourage you to see and act on a future that recognises that the life, safety and welfare of workers are sacred, and must be revered that way. Decision makers throughout society, especially in industry and workplaces, must put safety at the heart of their deliberations and decision making.

The events of Friday 25 September 1998 at the Esso Longford gas plant and the aftermath faced by the workers and their families must be written in the chronicles of Australian labour history, lest they be forgotten and repeated.

AFTER-NOTE TO THE READER

Thank you for reading my book. I hope it inspires you to live fairly and justly. And that no matter what your role in life is, when you are faced with a decision, treat others as you would want to be treated. Test your motives and make your mantra respect and dignity. If you use that measure to determine your course of action, you are bound to make the right decision.

I encourage you to join the My Life Matters Movement because when you know and believe that your life matters, you will take responsibility for defending and protecting it.

APPENDICES

October, 1999

"I'M NOT SURE WHERE WE'D BE TODAY IF WE HADN'T HAD THE UNION. NOT SURE AT ALL."

George Parker; Esso, Longford; Union Delegate, Australian Workers Union, Victorian Branch.

My name is George Parker. I started to work for Esso twenty-one years ago at thirty-five years of age after various efforts at market gardening and small business ventures.

In our country town I guess Esso employees have always done somewhat better than the average [person], so there were always those who wanted to work for the company.

I call myself a "new unionist" to distinguish myself from those people "born into the faith" and those such as myself who had never been involved in any form of unionism until a conversion later in life, so to speak.

When the union delegate was promoted in the middle of 1996, no one else was willing to accept the responsibility of taking over. So I offered to do the job for a year or so until a few of the boys were ready to take over. That year now has stretched to 18 months and in-between we have had a hectic and troublesome year.

Having swapped the shifts of that weekend with Ron, my wife Judy and I decided to have our own fishing trip with another couple.

The initial plans were to spend the night of Friday 25 September in a motel and continue with a relaxing and quiet weekend. But that whole day I felt "strange". I just can't explain it in any other way other than to say I felt strange.

Just about the time it [the explosion] happened we pulled up at a motel to check-in for the night. We looked around and talked about staying but suddenly I changed my mind and after a few words about better accommodation at home, virtually drove off without listening to the wishes of the others who wanted to stay. I don't know why I did [that].

After this, the mood in the car was far from the jovial and cheery mood you'd expect from a group of friends on a fishing weekend. I can only explain it as "strangely quiet". However, it got worse when our daughter Jane called on the mobile phone and told me that there had been "a huge problem at the plant". As happens with these incidents, the initial information was sketchy and mostly based on hearsay — adding to the concern and uncertainty I felt.

I don't recall any feelings at all. All I can say is that I felt "numb". I can't even remember if anyone said anything in the car or the time it took to get home after the telephone call.

I think these were the worst couple of hours of my life.

We always knew that we worked in an industry where that sort of thing could happen. But we also thought that it wouldn't happen to us. I really believed that at the end of the day if everything else failed, the shut down and pressure safety valve would save us. It didn't. That's what shocked me.

I felt a knot at the bottom of my gut knowing that my own shift was on. For the whole time I just thought, *oh God if my boys* [reference to the men on his shift] *were in that fire and if six were missing* — which was the initial number of workers thought to have been missing — which *six they would be?* There were maintenance people and managers and so the six could include some of them. But the law of averages says that we are the ones running the plant so some of us were going to be there. I didn't know what part of the plant the problem was in at that stage. All I knew was that it was in the plant.

Strangely enough, I don't think my initial thoughts were that I could have been there. I don't think that came into it. I just knew the boys from my shift would be affected and that's who I was worried about.

Anyhow, we just continued our way home. There was nothing I could do there.

As soon as I got home I changed my clothes and went straight out to the plant. I can't remember the details of those few hours, but I must have called some people because I knew that the boys had been evacuated and taken to St John's hall in Sale. Also, I had arranged to meet Terry the union organiser there. When I got there, they had all gone.

From there Terry and I went out to the plant and met with the managers from Melbourne and talked about where we were going to go and what was going to happen.

That was such an emotional time and so were the weeks to follow. A few times I said things that I probably didn't mean. For instance, on the night of the explosions one of the senior managers that I have known for years made the statement that he expected people who work for Esso to turn up immediately to help rebuild the plant. I really think he hadn't thought much about what he just said or the trauma the blokes must have been going through. I quickly told him that he would get the ones that could turn up and those who wanted to work, and they would be enough.

I guess it's changed most of our lives to some extent since then. Not all the experiences have been bad. I think it has been a learning experience for a lot of us. Certainly, I have learnt a lot from this experience. From that day on, I had to develop skills I never knew I had. Up until then, I thought I was one of the boys and I guess I think I always felt that I show some care for other people, but I never needed to show the understanding. Just sit, listen, comfort, talk, and let the boys do their own thing and air it out of their system.

Until it is necessary, you don't use those skills. I hope I have been some help to them over the last year or so. Most of them assured me I have. Some of them made the kind comments about me being the pillar that held them up. I don't know about that, but I certainly hope it's true. I only really did what anyone in my position would have done. I mostly sat with them and gave them moral support.

I also learnt some things about unionism. It is not about excluding people. It is about including everyone. The new leadership of Bill Shorten, the secretary, helped me reach out to those workers who had been isolated because they had disagreed with the union in one way or another. Now they are just as loyal as those who were union members for years.

I communicated with Billy Shorten daily and even several times a day when I needed to. I think that gave the blokes confidence in the knowledge that I represented them and their interest during meetings and interviews with the Victorian police, various government departments, the arson squad, company representatives, the WorkCover Authority, and the barrage of lawyers. Not that any one of the people from these organisations was hostile to them. To the contrary, they were very professional and respectful of the men, but the process was so new and intimidating that they just needed assurance.

Eventually, of course, they had to face the ordeal of giving evidence at the Longford Royal Commission hearings. It is sad, but I felt that the union was the only ray of hope most of them had during those harrowing times.

The assistance from the union in the past twelve months has been just outstanding. Terry Lee, our local organiser, has been terrific, particularly in the first week. Terry and another organiser, Sam Beechy, spent the entire week at the plant and gave us a lot of help, guidance and moral support. That has continued ever since.

The legal representation during the Royal Commission is something I would've never realised we needed and I don't believe many other workers anywhere would either. Bill Shorten realised the need and promptly organised legal assistance for us. I think the heroes in the legal aspect of this whole incident were Bill Shorten and the union lawyer, Bernard Murphy. They had outstanding and incredible foresight and skills.

We didn't realise what a Royal Commission meant. We had no idea we'd be involved in it. During the interviews conducted by the company representatives who'd been flown out from America to investigate the incident that I attended, they were saying, "You've got nothing to worry about. You did nothing wrong. You fellows did a wonderful job. It was a runaway train, and you had no control." So, why would we think we had anything to worry about? In view of what the company's investigating team had said to the people that were interviewed, there was just no thought of the workers being implicated in the way they eventually were.

As time went on, however, it became apparent just how vital the union's support and legal assistance was. Some patterns started to emerge during the Royal Commission. The way they cross-examined some of the boys on the stand made some of them feel that they were given a hard time. But that also led us to believe that they were aiming at blaming the boys, but it never came out that strongly.

I'm not sure where we'd be today if we hadn't had the union support us.

I think we wouldn't have known we were in trouble until we were [in trouble]. And we would have had no defence for the accusations the company made in their final submission. I'm sure the way they did their examining and cross-examining must have been a little more guarded too because they knew the union's barristers were there to cross-examine their witnesses and what have you.

And then they had a so-called independent expert they brought from the States. His report said that Esso had done everything perfectly. And it was the people they had working for them who knew what to do, but just didn't do it.

The union's lawyer, Bernard Murphy, did a great job of discrediting him very well. One of the questions he was asked was how many times he had worked for Esso or Exxon. In his statement, he'd indicated only four jobs. But our team had done their homework and knew it was many more. One by one our lawyer highlighted the other jobs he'd neglected to refer to in his statement.

Twenty jobs later, his credibility was looking shaky. Then they said, well you have done all these jobs for Exxon and Esso. Have you ever found any liability on their behalf? In the transcript, there's almost a page of dodging, etc. Then the commissioner ordered him to answer. He said no. So that was interesting.

It's been a difficult time. I mean these guys did everything they knew, and they did it to their best of their ability. They did far more on the day than they ever could have been expected to.

I always believe that when the crap hits the fan some of us will fail. Some will turn tail and run. It might be me, it might be someone else. That to me is how we're made. I thought somebody would fail no matter what training they'd had. But when I sat through the interviews with the arson squad with nearly everyone one of the boys, I realised none of them failed. None of them ran away and left their mates.

Sitting with them during those interviews I just started adding it all up and thought there wasn't one of those guys that did less than was required of them. There were a lot of them that did a hell of a lot more. But there was not one of them that did less than what their job required. It wasn't what each one said about himself, it was what each man said about the heroic efforts of another. And that made me proud to work with those guys.

It says a lot, because the human instinct says flight or fight and they all fought. Possibly some to different extents, but nobody did less than was expected of them. And some guys that'd been there for less than six months did fantastic things. I think it's more just the people they are.

The company absolutely knows that what happened wasn't the fault of those people, it wasn't the fault of Jim or Ronnie. If it were, then every other operator would've been at fault too because I have spoken to everyone and not one person has ever said to me, "I would have known what to do."

I think the right thing for Esso to do would have been to say that they didn't know how or what caused the explosions and fire. Instead they blamed Jim and to some extent Ronnie for what had happened.

There were at least nine people that were above Jim Ward in status on the day. There were supervisors, plant supervisors, and maintenance supervisors, some with 30 years' experience; why didn't they blame them?

The lawyer representing the union said that he thought it was quite simply Jim's evidence that was so clear and so damning that they were out to discredit him. Jim only told the truth. He is articulate and told it very well, very graphically, and explained himself very well. In my opinion, that's why they attacked him. I am sure of that. But there was no need to do that and there was no gain in it for anybody I don't think. They may have slightly reduced their liability by blaming somebody. That's all they ever hoped to do. Is that worth buggering up somebody's life for? I don't think so. I don't know, why else would they do it? It's just so hard for us to understand.

The day Esso brought out their final submission and blamed Jim and Ronnie was just too much for us. I think that's done some of the workers more harm than the accident. I think it's harder for them to get over that than the accident. To accept that the company blamed them for the tragedy on top of everything else they had suffered was too much. That was the time I got angry at them most. I really got out of control that night. That was the day, the only time I really lost it.

It was a terrible night. I think it was about 6 o'clock when I heard it. It was raining, and I remember driving to the plant and I talked to Bill Shorten on the mobile all the way. By this time the reconstruction had started, and the construction crew had just knocked off. There was just a continuous line of cars coming the other way and I was talking on the phone while I waited in my car in the rain.

By the time I got out there I was that worked up. I didn't know that one of the bosses from Melbourne was there and I just got out and said, "Haven't you f'd enough lives yet?" There were three of the managers there and they just sat there with their mouths open and they never said a word. And I probably paid out on them for five or ten minutes.

Probably four or five months later something happened, and I said to the same manager, "I am not going to make a fool of myself today." He asked what I meant, and I referred to this incident. He said, "Don't you ever be ashamed of that. You didn't make a fool of yourself and don't you ever be ashamed of it."

I'm not sure that the local management feels different than the rest of us. I get the feeling, and I am quite sure that at times, our plant manager particularly would love to express himself but he's not able to do so. I really think he is in a lot of ways a compassionate man, but he can only do what he is allowed to do.

They're certainly not saying anything and they're obviously not going to say anything. However, I feel that once you go above the Longford managers, the feeling may be that if you ignore it long enough it will go away. Why then won't they come over and meet with us?

I don't know of any occasion when the manager of the company has ever spoken to any of the workers at Longford before or after the accident. He's been down a few times, he's talked to management, he's done a plant tour a couple of times. But I don't know if he's ever walked into a control room or a work area and sat down and talked with the blokes. I don't believe he ever has. I think it is important. He is the leader of the team; he's the man making the decisions. He should be able to talk to the people he's making them about. He just hasn't.

I had a lot of trouble understanding how people could have a strong view that they believed was right and not express it. Or, go against that view to earn their wages at the end of the week. I almost think I'd rather bloody well starve than not say what I believed in.

I remember Jim asking one of the senior managers one night about the way people were treated and company policy and so on. After some discussion, the manager said to Jim, "When I signed on with the company I signed on for the full package and what's their view is my view."

You almost feel sorry for people that are like that. Bound. But on the other hand, when you look at it, I guess these people don't get to the positions they're in unless they are like that. I think it makes it a little bit easier to understand the problems we've got, not just in Esso but all around.

Internally, overall, I have handled it well. I have at times been very down. But I am lucky enough to have the ability to stand back just say that it is not good for me to go down with the blokes. I had a job to do.

I had to step away from it because if I was emotionally as bad as they were then I was not going to be able to do my job or be of any help to anybody. While the counsellors tried to help, the guys were saying they got more out of sitting down with me in half an hour than they did five hours with the counsellor. So, I used to sit down with them for half an hour. Sometimes I think some of the boys aren't right, which means that I haven't done it all that well.

Esso now has to face the consequences of its decision in many ways. It has driven loyalty of the workers towards the union, but it has also lost support in the community.

There are a lot of the workers that were in the union because they had to be, and I believe they are now true unionists. There's a huge difference now because they want to be in it and that's one of the good effects of what Esso did.

I find a lot of the blokes from the other shifts want to come and talk to me, whereas twelve months ago they wouldn't have done so. Now they come up and ask how things are going and catch up on union issues. I often hear them use 'we': we this, we that and are we doing such and such. There is a sense of unity that was not there before the incident. I think this is encouraging.

It is up to us to see that the support for the union never changes. It is a big job. I guess places like ours have got various groups and there'll always be those who want to do things differently.

As far the opinion of the community is concerned, I think I can give a good picture because I have lived in the community with a population of 16,000 for about 23-24 years. I know a lot more people now than I did before the disaster. But I knew a lot of the Longford community through my children's school council and so on. When the kids left school about ten years ago, I took a little holiday from such commitments. Anyhow, before the accident I think Esso held a reasonable position in town, especially as a preferred employer. That's changed a lot. Quickly.

A good example of the change of the town peoples' perception of Esso was given by a manager who has a lot to do with the local users. He probably talks to forty or fifty landholders on average a week. These include farmers, householders, shopkeepers and occupiers; a fair cross-section of the community. He said in the weeks after it had happened, there was a lot of sympathy for Esso as a group. For the people and the company. Within a day of Esso blaming Jim Ward, people would say to him, "You still work for those bastards?" He said it just changed like that. The dislike for the company in the local community was just incredible.

And they were vocal about it; very vocal. He said it wasn't one place; it was just incredible, just like that.

In my opinion, it was the worst decision the company may have ever made. But I think they don't really care about their image because they don't have to sell anything directly to the public. The consumers do not have a choice of going somewhere else.

Up until now they have a monopoly in this State and they have lost this position. Not because of the way they treated the workers but because they were thought to be unreliable.

The assistance from the union in the past twelve months for me as shop steward and in general for the members has been just outstanding. Probably due to one or two men in our case — namely Bill Shorten, the union secretary. That has continued ever since.

I am not sure where we'd be today if we hadn't had the union. Not sure at all.

FOOTNOTES

INTRODUCTION

1 Victoria, Royal Commission into Esso Longford Explosion, (1998), extract of letter dated 20 October 1998, of the Governor of Victoria the Honourable Sir James Gobbo AC appointing The Honourable Sir Daryl Michael Dawson & Brian John Brooks as the Commissioners to the Longford Royal Commission.

CHAPTER TWENTY
The Longford Royal Commission

2 Victoria, Royal Commission into Esso Longford Explosion, (1998), Vol 1, 6 [Preliminary Hearing]
3 Victoria, Royal Commission into Esso Longford Explosion, Report [4.4.1]
4 Ibid 7.7.2
5 Ibid 7.7.13
6 Ibid 7.7.19
7 Ibid 7.7.33
8 Ibid 7.7.68
9 Victoria, Royal Commission into Esso Longford Explosion, Report [1999], 236
10 Victoria, Royal Commission into Esso Longford Explosion, Report [1999], 234

ABOUT THE AUTHOR

Ramsina Lee has been working in the people side of businesses for over twenty years. She maintains that people are not 'resources'. People are human beings whose lives, safety and welfare are sacred and matter, and should be revered.

She attained her Bachelor of Arts degree with a double major in Industrial Relations and Sociology at the University of New South Wales before commencing her career, which includes serving on boards, tripartite councils and committees, as well as executive leadership teams across the private and public sectors, the unions and not-for-profit. She has contributed enormously to many organisations with her wealth of knowledge in all aspects of employee relations including leading the strategies for industrial relations, employee engagement, Work Health & Safety and across the broad range of people-related disciplines.

Ramsina believes that the measure to which society can progress, grow and prosper is the measure by which it respects the fundamental human rights of people to perform work that is safe, and in safe workplaces and their right to belong and associate.

She says that many people suffer needlessly, and unintentionally allow exploitation and hurt into their lives because of the lack of respect for life generally and because they do not believe that their life matters.

Ramsina started My Life Matters Movement to reverse this. Go to www.mylifemattersmovement.com for information about how to change the world by changing your belief that *your life matters.*

ACKNOWLEDGEMENTS

Where to start? An author may only have one opportunity to publicly thank all the people who have supported her and contributed to her work. So, it's a bit daunting to think I may have left someone out inadvertently. If I have, please forgive me.

Thank you to the following people.

To the workers and their families from Esso Longford who agreed to interview with me and to those who for their own reasons declined. This is your story. This is your voice. Know that your suffering has not gone unnoticed. I want this book to be included in the chronicles of Australian labour history as a reminder that lives matter and people matter.

To the Executive and members of the AWU for your assistance and support. With special thanks to Daniel Walton, National Secretary of the AWU and Ben Davis, Victorian State Secretary of the AWU for believing in my project and providing support.

The Honourable Mr Bill Shorten MP, Leader of the Australian Labor Party, the Opposition in Government at the time of printing of this book. Despite my anonymity as an author and your hectic schedule in public life, you said yes to my interview request because you have a heart for workers and you were so connected to the people and their families affected by the explosion and the following harrowing months leading up to and during the Longford Royal Commission. What would have become of them during those horrendous months and after, if you had not led their cause with your insight, skill and heart full of compassion? In 1998, you did not have the public profile you have now. Therefore, the intent of your actions can only be judged as noble and honourable.

Ian McDonald, Special Counsel, Simpsons Solicitors. Julie Postance, iinspire media. Sophie White, Creative Designer, the team at The Expert Editor, Tom Hyde, Liaison Librarian (Arts), University of Melbourne and Christina Ward, Librarian, Law Library University of Melbourne.

On a personal note.

To my mother. I have so much to thank you for. When I think of you, I think love and sacrifice. You loved us so much that you sacrificed your own pleasures, family connections, comforts and, at times, human dignity to make sure me, my brother and sister have a bright future. You are an amazingly strong and resilient woman. You inspire me to keep going no matter what obstacles come my way.

My husband Michael. You are the wind beneath my wings. I recall when I first shared my passion for writing and storytelling with you when we got married. You smiled, hugged me and declared, "I married an author!" That day, I felt encouraged to pursue my dream of writing. You encouraged me, accompanied me to the interviews, cried with me, felt incensed with me and shared this journey every step of the way. Finally, you cheered me on to publish it. This book is our book. Our daughters, Liz and Bec, if I were to state what I am most proud of and grateful to God for in my entire life, it is you two. You define joy and life itself for me and your father. You are the very breath I breathe. My heart leaped for joy the day I was told we were pregnant with you, and it is still leaping for joy. Thank you, my dear family, for agreeing to make the financial and time sacrifices to fund and complete this project. I owe you!

Sargon, Sam my brother, and Ramena, my sister. Having you two as my siblings has been a privilege and an honour. You have provided me with safety, security and comfort in my life since our childhood; knowing no matter what was happening around me, I could come to you for comfort, support and guidance. I trust you with my life. Anthony, my nephew, and Bianca and Catherine, my nieces, when you came into my life you moved into the centre of my universe. I loved you then and I love you now. You complete my picture of a happy family.

Kath Prendergast, my faithful and true friend. No words can describe the calm, wisdom and solidity you have brought into my once hectic life. Where on earth would I be without you? Sr Margaret Smith. You mean so much to me. You taught me what true, unconditional and non-judgemental love means by loving me so. You taught me that the intensity of my faith is in how I love myself and others, not how harshly I judge and am judged. Your joy is infectious and your unconditional love has been my life support when I've needed one.

www.ingramcontent.com/pod-product-compliance
Lightning Source LLC
Chambersburg PA
CBHW071918290426
44110CB00013B/1398